Lecture Notes in Mathematics

2217

More information about this series at http://www.springer.com/series/304

Sergey Bezuglyi • Palle E. T. Jorgensen

Transfer Operators, Endomorphisms, and Measurable Partitions

 Springer

Sergey Bezuglyi
Department of Mathematics
University of Iowa
Iowa City
Iowa, USA

Palle E. T. Jorgensen
Department of Mathematics
University of Iowa
Iowa City
Iowa, USA

ISSN 0075-8434   ISSN 1617-9692   (electronic)
Lecture Notes in Mathematics
ISBN 978-3-319-92416-8   ISBN 978-3-319-92417-5   (eBook)
https://doi.org/10.1007/978-3-319-92417-5

Library of Congress Control Number: 2018944127

Mathematics Subject Classification (2010): Primary: 47B65, 37B45, 28D05; Secondary: 37C30, 82C41, 28D15

Printed on acid-free paper

This Springer imprint is published by the registered company Springer International Publishing AG part of Springer Nature.
The registered company address is: Gewerbestrasse 11, 6330 Cham, Switzerland

Preface

While the fundamental notions, transfer operators, and Laplacians over decades have played a crucial role in diverse applications, their mutual interconnections still deserve a systematic study.

We begin by presenting a suitable measure theoretic framework for such a study of transfer operators, arising as part of the wider context of positive operators in measure spaces. From a long list of applications, we shall stress a detailed analysis of endomorphisms, of measurable partitions, and of Markov processes. Indeed, a number of earlier studies cover a host of special cases. Papers of special relevance to our present book include Vershik [VF85], [Ver00], [Ver01], [Ver05]; Baladi et al. [BER89], [Bal00], [BB05]; the first named author et al., e.g., [BKMS10], [BK16]; Alpay and the second named author, e.g., [AJL13], [AJK15], [AJLM15]; and Hawkins et al., e.g., [HS91], [Haw94].

Transfer operators arise in dynamical systems, and in a number of related applications; and a choice of transfer operator may serve different purpose from one application to the next. But typically a transfer operator serves to encode information about an iterated map, or iterated substitution systems. Here our focus will be an endomorphism in a fixed measure space and their iterations. Transfer operators are used in order to study the behavior of associated dynamical systems, and they occur in for example stochastic processes, in statistical mechanics, pioneered by David Ruelle, in quantum chaos, and in the study of such fractals that arise in iterated function systems (IFS). Each of these cases will be studied inside the book. Since transfer operators arise in multiple contexts, they also go by other names, for example *Ruelle operators* (after David Ruelle) or *Ruelle-Perron-Frobenius operators*. They are infinite-dimensional analogues of positive matrices and the Frobenius-Perron theorem. As in the matrix case, key questions are the determination of the eigenvalues, the spectral radius, and Perron-Frobenius eigenvector space. The role of eigenvalues and spectrum is also important in infinite dimensions, but then questions about spectrum are more subtle.

From our analysis of transfer operators, we pass to their connections to the theory of Laplace operators. The traditional setting of Laplacians ranges from the discrete domain to such measure theoretic frameworks that arise naturally in potential theory and in many diverse applications. The point of view of Laplacians has been studied in earlier papers, for example [JP11, JP13, JT16b, JT16a]. In the present book, we have stressed the "transfer operator point of view."

Our approach to transfer operators has been motivated by connections to recent studies of Laplace operators. Starting with the discrete case, recall that a discrete Laplace operator is an analogue of the better known continuous PDE-Laplace operator, the discrete variants arising from suitable discretization algorithms. They are used extensively in numerical analysis. A discrete Laplace operator is typically defined so that it has meaning on a graph G, i.e., a discrete grid of vertices and edges. Our present framework is the case when G is infinite, even possibly a measure space. In the various applications discussed in the book, we stress the connection to transfer operator theory.

For the case of finite graphs (G having a finite number of edges and vertices), the associated discrete Laplace operators are also called Laplacian matrices. In interesting applications, the analysis of infinite G may be achieved as suitable limits of the finite cases. However, recently Laplace operators have played an increasingly important role outside the setting of PDE theory.

Applications of Laplacians occur in physics problems, such as in Ising models, in loop quantum gravity, and in the study of discrete dynamical systems. Other applications include image processing, in the form of Laplace filters, and in machine learning for clustering and semi-supervised learning on neighborhood graphs. In addition to considering the connectivity of vertices (nodes) and edges in a graph, mesh Laplace operators take into account the underlying geometry (e.g., vertex angles). Different discretizations exist, some of them are an extension of the graph operator, while other approaches are based on the finite element method and allow for higher order approximations.

Iowa City, IA, USA   Sergey Bezuglyi
Iowa City, IA, USA   Palle E. T. Jorgensen
April 2018

Acknowledgments

The first named author is thankful to Professors Jane Hawkins, Olena Karpel, Konstantin Medynets, and Cesar Silva for useful discussions on properties of endomorphisms. The second named author gratefully acknowledges discussions, on the subject of this book, with his colleagues, especially helpful insight from Professors Daniel Alpay, Dorin Dutkay, Judy Packer, Erin Pearse, Myung-Sin Song, and Feng Tian.

The authors thank the members in the Operator Theory and Mathematical Physics seminars at the University of Iowa for enlightening discussions.

The authors are grateful to several anonymous reviewers for kindly preparing lists of corrections and for making constructive suggestions. We have followed them all. The book is better for it. Remaining flaws are the responsibility of the authors.

Contents

Chapter 1
Introduction and Examples

Abstract We present a unified study a class of positive operators called (generalized) transfer operators, and of their applications to the study of endomorphisms, measurable partitions, and Markov processes, as they arise in diverse settings. We begin with the setting of dynamics in standard Borel, and measure, spaces.

Keywords Positive operators · Endomorphisms · Transfer operators · Measurable partitions · Markov processes

While the mathematical structures of positive operators, endomorphisms, transfer operators, measurable partitions, and Markov processes arise in a host of settings, both pure and applied, we propose here a unified study. This is the general setting of dynamics in standard Borel and measure spaces. Hence the corresponding linear structures are infinite-dimensional. Nonetheless, we prove a number of analogues of the more familiar finite-dimensional settings, for example, the Perron-Frobenius theorem for positive matrices, and the corresponding Markov chains.

We develop a new duality between endomorphisms σ of measure spaces $(X, \mathcal{B})$, on the one hand, and a certain family of positive operators R acting in spaces of measurable functions on $(X, \mathcal{B})$, on the other. A framework of standard Borel spaces $(X, \mathcal{B})$ is adopted; and this generality is wide enough to cover a host of applications.

While the case of automorphisms in measure spaces has been extensively covered in the existence literature on ergodic theory and measurable dynamics, the study of endomorphisms (precise definitions below) is of more recent vintage. The latter is our focus on here: We show that a systematic study of endomorphisms dictates a whole new set of tools, often quite different from those used earlier for the "easier" case of automorphisms (in measure spaces). In the book, we identify, among other things, a family of positive operators (transfer operators) which arise naturally as a dual picture to that of endomorphisms. Our setting for positive operators is close to that of [Kar59]. Moreover, our approach to both the endomorphisms, and the associated transfer operators, is motivated by

© Springer International Publishing AG, part of Springer Nature 2018

S. Bezuglyi, P. E. T. Jorgensen, *Transfer Operators, Endomorphisms, and Measurable Partitions*, Lecture Notes in Mathematics 2217, https://doi.org/10.1007/978-3-319-92417-5_1

a number of recent applications. A more systematic list is given below; but the list includes the following: wavelets, more general classes of multi-resolution analyses (see e.g., [BFMP09, BMPR12, DJ06, Jor01, Sil13]), dissipative dynamical systems, and quantum theory. A parallel distinction from measurable dynamics, and thermodynamics, is that of time-reversible processes, as opposed to irreversible.

It is worth noting that the relationship of automorphisms to endomorphisms has parallels in operator theory; where the distinction is between unitary operators in Hilbert space versus such more general classes of operators as contractions. There is also a non-commutative version of automorphisms versus endomorphisms: While the study of automorphisms of von Neumann algebras dates back to von Neumann, the systematic study of endomorphisms (of von Neumann algebras) is more recent; see e.g., [AA01, BJP96, Jon94, Lon89, LP13, Mai13, Pow99, PP93].

In detail, from a given pair (R, σ) on $(X, \mathcal{B})$, a positive operator R, and an endomorphism σ, we define the notion of transfer operator. At the outset, measures on $(X, \mathcal{B})$ are not specified, but they will be introduced, and adapted to the questions at hand; in fact, a number of convex sets of measures on $(X, \mathcal{B})$ will be analyzed in order for us to make precise the desired duality correspondences between the two parts, operator and endomorphism, in a fixed transfer operator pair (R, σ). The theorems we obtain in this setting are motivated in part by recent papers dealing with stochastic processes (especially in joint work between D. Alpay et al. and the second named author), applications to physics, to path-space analysis, to ergodic theory, and to dynamical systems and fractals. A source of inspiration is a desire to find an infinite-dimensional setting for the classical Perron-Frobenius theorem for positive matrices, and for the corresponding infinite Markov chains. Indeed, recent applications dictate a need for such infinite-dimensional extensions.

Tools from the theory of operators in Hilbert space of special significance to us will be the use of a certain universal Hilbert space, as well as classes of operators in it, directly related to the central theme of duality for transfer operators. From ergodic theory, we address such questions as measurable cross sections, partitions, and Rohlin analysis of endomorphisms of measure spaces. While there are classical theorems dealing with analogous questions for *automorphisms* of measure spaces, a systematic study of *endomorphisms* is of more recent vintage;– in its infancy. In order to make the exposition accessible to students and to researchers in neighboring areas, we have included a number of explicit examples and applications.

The notion of transfer operators includes settings from statistical mechanics where they are often referred to as *Ruelle operators* (and we shall use the notation R for transfer operator for that reason), from harmonic analysis, including spectral analysis of wavelets, from ergodic theory of endomorphisms in measure spaces, Markov random walk models, transition processes in general; and more. The terminology "transfer operator" is from statistical mechanics; used for example in consideration of the action of a dynamical system on mass densities. The idea is that for chaotic systems, it is not possible to predict individual "atoms", or molecules, only the density of large collections of initial conditions [Rue78]. Or in mathematical language, "transfer operator" refers to the transformation of individual probability distributions for systems of random variables. There are further a number of parallels between our present infinite-dimensional theory and the classical Perron-Frobenius theorem for the special case of finite positive matrices.

To make the latter parallel especially striking, it is helpful to restrict the comparison to the case of the Perron-Frobenius for finite matrices in the special case when the spectral radius is 1 (see e.g., [Bal00, BB05, BJL96, Kea72, MU15, NR82, Par69, Rad99]).

As we hint at in the title to our book, in our infinite-dimensional version of Perron-Frobenius transfer operators, we include theorems which may be viewed as analogues of many points from the classical finite-dimensional Perron-Frobenius case, for example, the classical respective left and right Perron-Frobenius eigenvectors, now take the form in infinite-dimensions of a *positive R-invariant measure* (left) and the infinite-dimensional right Perron-Frobenius vector becomes a *positive harmonic function*.

Of course, in infinite-dimensions, we have more non-uniqueness than is implied by the classical matrix version, but we also have many parallels. We even prove infinite-dimensional versions of the Perron-Frobenius limit theorem from the classical matrix case.

In recent research (detailed citations below) in infinite-dimensional analysis, a number of frameworks have emerged that involve positive operators, but nonetheless, a *unified* infinite-dimensional setting is only slowly taking shape. While these settings and applications involve researchers from diverse areas, and may on the surface appear quite different, they, in one way or the other, all involve generalizations of the classical Perron-Frobenius theory which in turn has already found many applications in ergodic theory, in the study of Markov chains, and more generally in infinite-dimensional dynamics.

Motivated by recent research, it is our aim here to address and unify these infinite-dimensional settings. Our work in turn is also motivated by many instances of the use of classes of positive operators which by now go under the name "transfer operators," or Ruelle operators, (see below for precise definitions). The latter name is after David Ruelle who first used such a class of these operators in the study of phase transition questions in statistical mechanics. Subsequent research on such questions as symbolic dynamics, spectral theory, endomorphisms in measure spaces, and diffusion processes, further suggest the need for a unifying infinite-dimensional approach. In fact the list of applications is longer than what we already hinted at, and it includes recent joint research involving the second named author with Daniel Alpay, and collaborators; details and citations are included below e.g., [AJL13, AJL16, AL13]). This collaborative research also makes use of positive operators and transfer operators in several infinite-dimensional settings, specifically in the study of such stochastic processes as infinite-dimensional Markov transition systems, analysis of Gaussian processes, and in the realization of wavelet multiresolution constructions for a host of probability spaces, and their associated L^2 Hilbert spaces, all of which go beyond the more traditional setting of $L^2(\mathbb{R}^d)$ from wavelet theory. Indeed the last mentioned multi-scale wavelet constructions are applicable to a general framework of self-similarity from geometric measure theory (see e.g., [Kea72, KFB16, Hut81, HR00]).

Important points in our present consideration of transfer operators are as follows: We formulate a general framework, a list of precise axioms, which includes a diverse host of cases of relevance to applications. In this, we separate consideration of the transfer operators as they act on *functions on Borel spaces* $(X, \mathcal{B})$ on the one hand, and their *Hilbert space properties* on the other. When a transfer operator is given, there is a variety of measures compatible with it, and we will discuss both the individual cases, as well as the way a given transfer operator is acting on a certain universal Hilbert space. The latter encompasses all possible probability measures on the given Borel space $(X, \mathcal{B})$. This approach is novel, and it helps us organize our discussion of a host of ergodic theoretic properties relevant to the theory of transfer operators.

The chapters in the book are organized as follows: The early chapters are in the most general setting, and the framework is restricted in the later more specialized chapters. Each specialization in turn is motivated by applications. To make the book accessible to a wider readership, including non-specialist, at the end of these chapters we have cited some papers/books which may help by discussing foundations, applications, and motivation.

A detailed summary of our main theorems is given in Sect. 1.4 below.

1.1   Motivation

This work is devoted to the study of *transfer operators*, see Definition 1.1. This kind of operators, acting in a functional space, has been studied in numerous research papers and books. They are also known by the name of *Ruelle operators* or *Perron-Frobenius operators* that are used synonymously. One of the first instances of the use of transfer operators in the sense we address here was papers by Ruelle in the 1970s (see, e.g., [Rue78]) dealing with phase transition in statistical mechanics. Since then the subject has branched off in a variety of new directions, and new applications. Our present aim is to give a systematic and general setting for the study of transfer operators, and to offer some key results that apply to this general setting. Nonetheless, by now, the literature dealing with transfer operators and their diverse applications is large. For readers interested in the many settings in dynamical systems where some version, or the other, of a transfer operator arises, we have cited the papers below [Kea72, Rue78, Rue89, BB05, BER89, BJL96, Bal00, Rue92, Rue02, Dut02, DR07, Jor01, Kat07, MU10, MU15, Rad99, Sto12, Sto13]. Non-singular transformations of measure spaces are of a special interest. We refer to the following papers in this connection [BG91, BH09, DH94, ES89, HS91, Sil88]. Invariant measures on Cantor sets are studied, in particular, in the following papers [BKMS10, BKMS13, BH14, BK16].

Our present results are motivated in part by *applications*. These applications include Markov random walk models, problems from statistical mechanics, and from dynamics. While our setting here, dealing with transfer operators and endomorphisms in general measure spaces, is of independent interest, there are also a number of more recent applications of this setting to problems dealing with generalized multi-resolution analysis, relevant to the study of wavelet filters which require the use of solenoid analysis for their realization. In fact, the following is only a sample of research papers devoted to these problems [BFMP09, BMPR12, FGKP16, LP13].

Since our work touches rather *different areas of Analysis*, we give here a list of principal references in the corresponding fields. While there is a rich literature on *endomorphisms* of non-Abelian algebras of operators, both C*-algebras, and von Neumann algebras, the nature of endomorphisms of Abelian measure spaces presents intriguing new questions which are quite different from those studied so far in the corresponding non-Abelian situations. Our present analysis deals with endomorphisms of Abelian measure spaces. (The interplay between the Abelian vs the non-Abelian case is at the heart of the Kadison-Singer problem/now theorem, see [CT16, MSS15], but this direction will not be addressed here.) For readers interested in the non-Abelian cases, we offer the following references [Lon89, BEK93, BJP96, BJ97, Jor01, BJ02, BK00, PP93, Pow99, Jon94, BKLR15]. The notion of a transfer operator for C^*-algebras was considered in [BJ97, Exe03].

The study of *transfer operators*, and more generally *positivity-preserving operators*, are both of independent interest in their own right. This is in addition to its use in numerous applications; both within mathematics, and in neighboring areas; for example in physics, in signal analysis, in probability, and in the study of stochastic processes. While we shall cite these applications inside the book, we already now call attention to the following recent papers [AJL13, AJL13, AK13, AL13, AJS14, AJV14, AJK15, AK15, AJ15, AJLM15, AJLV16, ACKS16].

While the notion of *resolution, multi-resolution,* or *multi-resolution analysis* has its roots in vision, or image analysis, a more recent use of the same idea is to a relatively new variety of wavelet constructions, or signal/image processing algorithms; see eg., [BJ02], and the cited papers below. Here we shall adapt a variant of this general idea to measurable dynamical systems. More precisely, our focus is on non-invertible endomorphisms of measure spaces. In some cases, the latter may be realized as subshifts in suitable choices of symbol space.

Our main theme will be to develop a systematic setting for, and approach to, the dynamics of endomorphisms in measure spaces. The idea is to introduce *encoding mappings* which will then in turn make direct connections to *wavelet* constructions and to analysis of *iterated function systems* (IFSs). In the case of wavelet resolutions, see e.g., [BJ02, AJL13, AJL16, BFMP09], the idea is that a choice of *scaling function*, say $\varphi \in L^2(\mathbb{R}^d)$, determines an initial resolution corresponding to the closed subspace in $L^2(\mathbb{R}^d)$; it is spanned by all discrete vector translates of the chosen scaling function φ, more precisely translates of φ by points from the integer grid $\mathbb{Z}^d$.

The next step in the wavelet resolution is scaling in $L^2(\mathbb{R}^d)$: A scaling operator typically acts on functions in $L^2(\mathbb{R}^d)$ by scaling with an expansive integer $d \times d$ matrix. Iteration of the corresponding scaling operator (realized as a unitary operator in $L^2(\mathbb{R}^d)$) will then lead to a resolution of $L^2(\mathbb{R}^d)$ into a scale of nested subspaces of the initial resolution subspace, the smaller subspaces corresponding to coarser resolutions. One of the many advantages of the approach is that it leads to efficient algorithms, called wavelet algorithms. And we note that the algorithms are built in symbol space, and then transferred to a wavelet construction in, for example, $L^2(\mathbb{R}^d)$.

We cite some papers on the *multiresolutions* that are related to our work [BJ02, BJMP05, KFB16, BRC16, SG16, AJLV16].

Iterated function systems (IFS) are used to describe the properties of fractal sets, and have close relations to transfer operators. Here we cite papers on IFS and their connections to various aspects of transfer operators: [BHS08, BHS12, Bea91, Hut81, Hut96, HR00, Rue78, Rue89, Rue92, Rue02, YLZ99]. We discuss IFSs and their relations to transfer operators in Chaps. 11–13. Also the concept of IFS plays a leading role in the set of examples which are considered below in this chapter. As proved in [Hut81], every IFS generates a uniquely defined probability measure μ (we call it IFS measure). The properties of μ is an important characteristic of the IFS. For example, the planar Sierpinski measure μ defined on the Sierpinski gasket admits a direct integral (disintegration) representation.

1.2 Examples of Transfer Operators

Our goal is to study *transfer operators* in the framework of various functional spaces. To be more specific, we briefly mention several typical examples of transfer operators. They will illustrate our results proved below. The rigorous definitions of used notions are given in the next chapter, see also Definitions 1.1 and 3.1.

Our approach to the theory of transfer operators can be briefly described as follows. We first define and study these operators in the most abstract setting, aiming to find out what general properties they have. By *abstract setting*, we mean the space of Borel real-valued functions $\mathcal{F}(X, \mathcal{B})$ over a *standard Borel space* (X, B). Such spaces being endowed with a topology, or a Borel measure, are used in most interesting classes of transfer operators such as Frobenius-Perron operators, or operators corresponding to iterated function systems, or operators acting in a Hilbert space, etc.

Let σ be a fixed *surjective Borel endomorphism* of $(X, \mathcal{B})$, and let $M(X)$ be the set of all Borel (finite or sigma-finite) measures on $(X, \mathcal{B})$. In general, $\sigma^{-1}(\mathcal{B}) := \{\sigma^{-1}(A) : A \in \mathcal{B}\}$ is a proper nontrivial subalgebra of $\mathcal{B}$ where $\sigma^{-1}(A) = \{x \in X : \sigma(x) \in A\}$. In fact, an endomorphism σ defines a sequence of filtered subalgebras $\mathcal{B} \supset \sigma^{-1}(\mathcal{B}) \supset \cdots \sigma^{-n}(\mathcal{B}) \supset \cdots$. An important property of σ, called *exactness*, is characterized by the triviality of the subalgebra $\mathcal{B}_\infty = \bigcap_{n \in \mathbb{N}} \sigma^{-n}(\mathcal{B})$, see Definition 2.5. We note that a Borel function f on $(X, \mathcal{B})$ is $\sigma^{-1}(\mathcal{B})$-measurable if and only if there exists a Borel function g such that $f = g \circ \sigma$.

When a measure $\lambda \in M(X)$ is fixed, then we get into the framework of a *standard measure space* $(X, \mathcal{B}, \lambda)$ (see, e.g., [CFS82]), and, in this situation, we use measurable sets from the complete sigma-algebra[1] $\mathcal{B}(\lambda)$ and functions measurable with respect to $\mathcal{B}(\lambda)$ instead of Borel ones. With some abuse of notation, we will also use the same symbol $\mathcal{B}$ for the sigma-algebra of measurable sets.

Having these data defined, we now give the following main definition.

Definition 1.1 Let $\sigma : X \to X$ be a surjective endomorphism of a standard Borel space $(X, \mathcal{B})$. We say that R is a *transfer operator* if $R : \mathcal{F}(X) \to \mathcal{F}(X)$ is a linear operator satisfying the properties:

(i) R is a positive operator, that is $f \geq 0 \implies Rf \geq 0$;
(ii) for any Borel functions $f, g \in \mathcal{F}(X)$,

$$R((f \circ \sigma)g) = f R(g). \tag{1.1}$$

If $R(\mathbf{1})(x) > 0$ for all $x \in X$, then we say that R is a *strict* transfer operator (here and below $\mathbf{1}$ means the constant function that takes value 1). If $R(\mathbf{1}) = \mathbf{1}$, then R is called a *normalized* transfer operator.

Relation (1.1) is called the *pull-out property*.

In what follows we describe several classes of transfer operators and then give a universal approach to these classes based on the notion of a *measurable partition*, see Sect. 2.1. More examples of transfer operators will be also given in subsequent chapters.

Example 1.2 (Transfer Operators Defined by Finite-to-One Endomorphisms) Let $X = [0, 1)$ be the unit interval with Lebesgue measure dx. Take the endomorphism σ of X into itself defined by

$$\sigma(x) = 2x \bmod 1.$$

Then σ is onto, and $|\{\sigma^{-1}(x) : x \in X\}| = 2$. Consider a function space $\mathcal{F}$ of real-valued functions over X. We do not need to specify this space here. For instance, it can be either $L^p(X, dx)$, or the space of all Borel functions, or the space of continuous functions, etc. Set

$$R_\sigma(f)(x) := \frac{1}{2}\left(f(\frac{x}{2}) + f(\frac{x+1}{2})\right), \quad f \in \mathcal{F}. \tag{1.2}$$

Relation (1.2) gives an example of a transfer operator that is well studied in the theory of iterated function systems (IFS).

Based on this elementary example, we can use a more general approach to the definition of R_σ. Suppose that σ is an n-to-one endomorphism of a measurable

---

[1] We reserve the symbol σ for an endomorphism of a standard Borel space $(X, \mathcal{B})$, so that to avoid any confusion we write sigma-algebra and sigma-finite measure instead of such more common terms σ-algebra and σ-finite.

Table 1.1 Invariant measures for R_σ and R'_σ

Transfer operator	Lebesgue measure μ	Dirac measure δ_0
R_σ	$\mu R_\sigma = \mu$	$\delta_0 R_\sigma = 1/2(\delta_0 + \delta_{1/2}) \not\ll \delta_0$
R'_σ	$\mu R'_\sigma \ll \mu, d(\mu R'_\sigma) = 2\cos^2(\pi x)d\mu(x)$	$\delta_0 R'_\sigma = \delta_0$

space $(X, \mathcal{B})$, and $\mathcal{F}(X)$ is an appropriate functional space of real-valued functions. Let W be a nonnegative function on X (it is called a weight function). We define a transfer operator $\mathcal{F}(X)$ by the formula

$$R_\sigma(f)(x) = \sum_{y \in \sigma^{-1}(x)} W(y)f(y). \qquad (1.3)$$

Clearly, $R_\sigma f \geq 0$ whenever $f \geq 0$, i.e., R_σ is a positive operator. Moreover, if **1** denotes the constant function that takes the value 1, then the condition $R_\sigma(\mathbf{1}) = \mathbf{1}$ holds if and only if $\sum_{y \in \sigma^{-1}(x)} W(y) = 1$ for all x. The most important fact about R_σ is that if R_σ satisfies the pull-out property: for any functions f and g from $\mathcal{F}(X, \mathcal{B})$,

$$R_\sigma((f \circ \sigma)g)(x) = f(x)(R_\sigma g)(x). \qquad (1.4)$$

In case of the transfer operator given in (1.2), it can be modified by considering a nontrivial weight function W. Illustrating our further results, we will deal with R_σ defined by (1.2), or more generally by

$$R'_\sigma(f)(x) := \cos^2\left(\frac{\pi x}{2}\right)f\left(\frac{x}{2}\right) + \sin^2\left(\frac{\pi x}{2}\right)f\left(\frac{x+1}{2}\right), \quad f \in \mathcal{F}, \qquad (1.5)$$

as well.

As we will see below, any normalized transfer operator defines an action on the set of probability measures. It is interesting to note that R_σ and R'_σ have different properties relating to the corresponding invariant measures. We present them in the following table. More detailed exposition of these results is given in Chap. 13. See (1.2) and (1.5) for the definition of operators R_σ and R'_σ (Table 1.1).

Remark that we used in the table the fact that transfer operators define the dual action on space of measures, see Chap. 3 for more details.

The following class of transfer operators is a continuous analogue of the operators defined by (1.3).

Example 1.3 (Frobenius-Perron Operators)   We follow here [AA01, LM94, DZ09]. Suppose we have a standard measure space $(X, \mathcal{B}, \mu)$ and a surjective non-singular endomorphism σ acting on the space $(X, \mathcal{B}, \mu)$. Let P be a positive operator on $L^1(X, \mathcal{B}, \mu) = L^1(\mu)$. It is said that P is a *Frobenius-Perron operator* if for any $f \in L^1(\mu)$, and any set $A \in \mathcal{B}$,

$$\int_A P(f) \, d\mu = \int_{\sigma^{-1}(A)} f \, d\mu. \qquad (1.6)$$

It can be easily checked that this Frobenius-Perron operator satisfies the pull-out property (1.1). Furthermore, it follows from (1.6) that P preserves the measure μ, i.e., $\mu P = \mu$ where μP is defined by the formula:

$$(\mu P)(A) = \int_X P(\chi_A)\,d\mu.$$

Relation (1.6) can be considered as a partial case of more general approach. We can define a "non-singular" Frobenius-Perron operator as follows:

$$\int_A P(f)\,d\mu = \int_{\sigma^{-1}(A)} Wf\,d\mu. \tag{1.7}$$

Then W can be viewed as the *Radon-Nikodym derivative* of the measure μP with respect to μ. Note that if $W = 1$, then the measure μ is P-invariant, and relation (1.7) is reduced to (1.6).

Example 1.4 (Transfer Operators on Densities) Let σ be an onto endomorphism of a standard Borel space $(X, \mathcal{B})$. Fix a Borel measure λ on $(X, \mathcal{B})$ such that $\lambda \circ \sigma^{-1} \ll \lambda$. Define a linear operator $R = R_\lambda$ acting on non-negative functions f from $L^1(\lambda)$ by the formula

$$R_\lambda(f)(x) = \frac{(f d\lambda) \circ \sigma^{-1}}{d\lambda}, \tag{1.8}$$

where the right-hand side is the Radon-Nikodym derivative of the measure $(f d\lambda) \circ \sigma^{-1}$ with respect to λ. Then R_λ is called a Ruelle transfer operator. It can be easily checked that R_λ satisfies the conditions of Definition 1.1: (i) R_λ is positive, (ii) $R_\lambda((f \circ \sigma)g) = f R_\lambda(g)$ for any $f, g \in L^1(\lambda)$. We note that this operator R_λ simultaneously acts on the set of Borel measures $M(X)$. The pull-out property of R_λ (1.1) can be written in the equivalent form

$$\int_X g(Rf)\,d\lambda = \int_X (g \circ \sigma)f\,d\lambda.$$

Then one sees that $\lambda R_\lambda = \lambda$.

It turns out that the transfer operators defined in Examples 1.3 and 1.4 are related in a simple way.

Lemma 1.5 *Let μ be a Borel measure on $(X, \mathcal{B})$, and let P be a Frobenius-Perron operator acting on $L^1(X, \mathcal{B}, \mu)$ as in (1.6). Then $P(f) = R_\mu(f)$ for $f \in L^1(\mu)$. If P is defined by (1.7), then $P(f) = R_\mu(Wf)$.*

Proof Indeed, relation (1.7) can be written in an equivalent form as

$$\int_X P(f)g\, d\mu = \int_X (g \circ \sigma)fW\, d\mu.$$

Then the lemma follows.   □

The next example is important and will be used later, see Chaps. 4 and 13.

Example 1.6 (Transfer Operators Via Systems of Conditional Measures)   This example of a transfer operator is of different nature and is based on the notion of a *system of conditional measures*. The definitions of used terms can be found in Chap. 2.

Let $(X, \mathcal{B}, \mu)$ be a standard measure space with finite measure, and let σ be an endomorphism onto X. Consider the *measurable partition* ξ of X into preimages of σ, $\xi = \{\sigma^{-1}(x) : x \in X\}$. Take the system of conditional measures $\{\mu_C\}_{C \in \xi}$ corresponding to the partition ξ (see Definition 2.7).

We define a transfer operator R on the standard probability measure space $(X, \mathcal{B}, \mu)$ by setting

$$R(f)(x) := \int_{C_x} f(y)\, d\mu_{C_x}(y) \qquad (1.9)$$

where C_x is the element of ξ containing x, i.e., $C_x = \sigma^{-1}(x)$. The domain of R is $L^1(\mu)$ in this example.

Lemma 1.7 *The operator $R : L^1(\mu) \to L^1(\mu)$ defined by (1.9) is a transfer operator.*

Proof Clearly, R is a positive operator. To see that (1.1) holds, we simply calculate

$$R((f \circ \sigma)g)(x) = \int_{C_x} f \circ \sigma(y)g(y)\, d\mu_{C_x}$$

$$= f(x) \int_{C_x} g(y)\, d\mu_{C_x}(y)$$

$$= f(x)(Rg)(x).$$

Here we used the fact that $f(\sigma(y)) = f(x)$ for $y \in C_x = \sigma^{-1}(x)$.   □

If the operator R, defined in (1.9), is considered on $L^1(\mu)$ or $L^2(\mu)$, then R is transformed to the conditional expectation. An abstract analogue of conditional expectation is contained in Corollary 3.15. More results about this type of transfer operators are discussed in Chap. 13.

1.3   Directions and Motivational Questions

In this section, we formulate, in a rather loose manner, a few problems that could be considered as directions of further work in this area.

If a transfer operator R is defined by an endomorphism σ, then, as we will see below, it is convenient to view at R as a pair (R, σ). This notation makes sense because the set of such pairs forms a semigroup, and moreover it emphasizes that these two objects are closely related to each other, according to the "pull-out property" given in (1.1) and (3.1). Next, this point of view is useful for the problem of classification of transfer operators. Clearly, the set $\mathcal{R}_\sigma$ of transfer operators defined by the same endomorphism σ can be vast, as we have seen in the examples given in this chapter.

To understand better the research directions of our approach, we mention here a few questions which are not rigorously formulated but nevertheless serve as motivational questions. Obviously, the study of possible relations between *Borel dynamical systems* $(X, \mathcal{B}, \sigma)$, or *measurable dynamical systems* $(X, \mathcal{B}, \mu, \sigma)$, and *transfer operators* R, is a big multifacet problem, and we do not try to discuss all aspects of it here.

In Detail

(A) Suppose an endomorphism σ is given in a standard Borel space $(X, \mathcal{B})$. Denote by $\mathcal{R}_\sigma$ the set of all transfer operators (R, σ) on $\mathcal{F}(X)$. What can be said about the properties of the set $\mathcal{R}_\sigma$? Clearly, $\mathcal{R}_\sigma$ is a convex set. How can one find its *extreme points*? This question becomes clearer when a measure μ is fixed on $(X, \mathcal{B})$ and the operators (R, σ) are considered in $L^p(\mu)$-spaces.

(B) The interaction between dynamical properties of endomorphisms and transfer operators, such as *ergodicity, mixing*, etc., has been discussed in many papers, see e.g., [LM94, DZ09]. Our main interest is the study of the set of measures which are quasi-invariant for both transformations, σ and R. This approach has been productive for the Frobenius-Perron operators defined in this chapter.

(C) In treating positive, and transfer operators, R as infinite-dimensional analogues of positive matrices, it is natural to raise the questions about *spectral properties* of such operators. If h is a harmonic function for R and a measure μ is invariant (or "quasi- invariant") with respect to R, then the relations $Rh = h$ and $\mu R = \mu$ ($\mu R \ll \mu$, respectively) are infinite dimensional analogues of eigenvectors in the matrix case for R. It would be interesting to find out how far the analogue with positive matrices can be extended to transfer operators.

(D) In the definition of a transfer operator, it is required that R is an operator defined on the set of functions. In some cases, this action generates a *"dual" action* of R on the set of all Borel measures $M(X)$. For instance, this is true for transfer operators defined on continuous functions over a compact Hausdorff space. How can one find, say, measures invariant with respect to R? Is there an interaction between actions of R on functions and on measures? In particular, we can define an equivalence relation on the set of all transfer operators. Given

(R, σ), let $\mathcal{I}(R, \sigma)$ be the set of all probability measures which are invariant with respect to R and σ. It is said that (R_1, σ_1) and (R_2, σ_2) are *measure equivalent* if $\mathcal{I}(R_1, \sigma_1) = \mathcal{I}(R_2, \sigma_2)$. How can transfer operators be classified with respect to the measure equivalence relation?

(E) We will study transfer operators R acting in various functional spaces. The same transfer operator R and endomorphism σ can be considered in different frameworks depending on the *choice of its domain*. For instance, if X is a compact Hausdorff space and σ is a continuous map on X, then it is natural to consider a transfer operator (R, σ) as acting on continuous functions $C(X)$. At the same time, (R, σ) can be viewed as a transfer operator on the space of Borel functions $\mathcal{F}(X, \mathcal{B})$, or on the space $L^p(X, \mathcal{B}, \lambda)$. It would be interesting to understand how properties of R depend on the choice of an underlying space.

1.4   Main Results

A common theme is as follows: Given a transfer operator (R, σ), what are the properties and the interplay between the following dual actions, action of R on *functions* vs its action on *measures*? What is the interplay between the action of R and that of an associated endomorphism σ? What are the important classes of quasi-invariant measures? These questions are answered in Chaps. 4–6, see especially Theorems 4.14, 4.19, 4.22, 5.9, 5.12, 6.6, and 6.9. We also mention our main Theorems 7.3, 7.5, 8.17, 11.4, 11.5, 11.15, 13.1, 13.6 from other chapters (more important results are obtained in Corollaries 9.4, and 12.7).

The necessary preparation and preliminary results are in Chaps. 2 and 3.

For each of the classes of *quasi-invariant* measures, when do we have existence? This is the Perron-Frobenius setting, and now made precise in the general infinite-dimensional setting, and involving harmonic functions and measurable partitions. Our answers here are in Theorems 5.20, 8.12, 8.18, and in Proposition 5.17.

When does a given transfer operator (R, σ) induce a multiresolution, i.e., a filtered system of subspaces, or of measures? And under what conditions does exactness hold? (See Theorem 6.9). This type of questions is discussed in Chaps. 3, 6, and 7.

In Theorems 4.19 and 10.6, we establish explicit *measurable partitions*, co-boundary analysis, ergodic properties, and ergodic decompositions. In Theorem 8.12, we show that there is a *universal Hilbert space* which realizes every transfer operator (R, σ).

Chapter 2
Endomorphisms and Measurable Partitions

Abstract In this chapter, we collect definitions and some basic facts about the underlying spaces, endomorphisms, measurable partitions, etc., which are used throughout the book. Though these notions are known in ergodic theory, we discuss them for the reader's convenience.

Keywords Endomorphism · Standard measure space · Measurable partitions · Radon-Nikodym derivative · Polish space · Sigma-algebra of Borel sets · Solenoids

2.1 Standard Borel and Standard Measure Spaces

Definition 2.1 Let X be a separable completely metrizable topological space (a *Polish space*, for short), and let $\mathcal{B}$ be the *sigma-algebra of Borel subsets* of X ($\mathcal{B}$ is the smallest sigma-algebra generated by open sets). Then $(X, \mathcal{B})$ is called a *standard Borel space*. If μ is a continuous (i.e., non-atomic) Borel measure on $(X, \mathcal{B})$, then $(X, \mathcal{B}, \mu)$ is called a *standard measure space*.

In this book, we will deal only with standard Borel and measure spaces (sometimes it is clearly formulated in statements). We use the same notation, $\mathcal{B}$, for Borel sets, and measurable sets, of a standard measure space. It will be clear from the context in what settings we are. Dealing with the sigma-algebra of measurable sets, we will assume that $\mathcal{B}$ is *complete* with respect to the measure μ. The set of all sigma-finite complete Borel positive measures on $(X, \mathcal{B})$ is denoted by $M(X)$. Let $M_1(X) \subset M(X)$ denote the subset of probability measures. For short, an element of $M(X)$ will be called a *measure*. If μ and ν are two measures from $M(X)$, then μ is absolutely continuous with respect to ν, $\mu \ll \nu$, if $\nu(A) = 0$ implies $\mu(A) = 0$. Two measures μ and ν on $(X, \mathcal{B})$ are called *equivalent*, $\mu \sim \nu$, if they share the same sets of measure zero, i.e., $\mu \ll \nu$ and $\nu \ll \mu$.

© Springer International Publishing AG, part of Springer Nature 2018
S. Bezuglyi, P. E. T. Jorgensen, *Transfer Operators, Endomorphisms, and Measurable Partitions*, Lecture Notes in Mathematics 2217,
https://doi.org/10.1007/978-3-319-92417-5_2

We denote by $\mathcal{F}(X)$ (or by $\mathcal{F}(X, \mathcal{B})$) the vector space of Borel real-valued functions. If a measure μ is defined on $(X, \mathcal{B})$, we will work with the space of measurable real-valued functions with respect to μ.

All objects considered in the context of measure spaces (such as sets, partitions, functions, transformations, etc) are considered by modulo sets of zero measure (they are also called null sets). In most cases, we will implicitly use this   mod 0 convention.

It is a well known fact that all uncountable standard Borel spaces are Borel isomorphic, and that all standard measure spaces are measure isomorphic. This means that results do not depend on a specific realization of an underlying space. We will discuss this issue in the context of *isomorphic transfer operators* in Chap. 3.

2.2   Endomorphisms of Measurable Spaces

The notion of an endomorphism is a central concept of ergodic theory and endomorphisms are studied extensively in many books and research papers. We mention only a few of them to present a wide spectrum of research directions: [Roh61], [Haw94], [CFS82], [Bén96], [BH09], [PU10], [Bog07], [Ver01], and [Ren87]. Rokhlin started a systematic study of measurable dynamical systems. His pioneering works on measurable partitions [Roh49b], and on properties of automorphisms and endomorphisms of a of a standard measure space (see [Roh49a, Roh61]) opened new directions in the ergodic theory. We recall his famous results, the *Rokhlin lemma*, which states that any given aperiodic measure preserving automorphism of a measure space can be approximated by a periodic transformation up to a set of arbitrarily small measure.

Let σ be a Borel map of $(X, \mathcal{B})$ onto itself. Such a map σ is called an onto *endomorphism* of $(X, \mathcal{B})$. In particular, σ may be injective; in this case, we have a Borel *automorphism* of $(X, \mathcal{B})$. Since the cardinality of the set $\sigma^{-1}(x)$ is a Borel function on X, we can independently consider the following classes: σ is either a finite-to-one or countable-to-one map, or $\sigma^{-1}(x)$ is an uncountable Borel subset for any $x \in X$. In general, we do not require that the set $\sigma(A)$ is Borel but if σ is at most countable-to-one, then this property holds automatically.

We denote by $End(X, \mathcal{B})$ the semigroup (with respect to the composition) of all surjective endomorphisms of the standard Borel space $(X, \mathcal{B})$.

Given an endomorphism σ of $(X, \mathcal{B})$, we denote by $\sigma^{-1}(\mathcal{B})$ the proper subalgebra of $\mathcal{B}$ consisting of sets $\sigma^{-1}(A)$ where A is any set from $\mathcal{B}$.

We will use endomorphisms mostly in the context of standard measure spaces $(X, \mathcal{B}, \mu)$ with a finite (or sigma-finite) measure μ. Any endomorphism σ of

$(X, \mathcal{B}, \mu)$ defines an action on the set of measures $M(X)$ by

$$\mu \mapsto \mu \circ \sigma^{-1} : M(X) \to M(X),$$

where $(\mu \circ \sigma^{-1})(A) := \mu(\sigma^{-1}(A))$. For a fixed measure μ, it is said that σ is a *non-singular endomorphism* (or equivalently that $\mu \in M(X)$ is a (backward) *quasi-invariant measure* with respect to σ) if $\mu \circ \sigma^{-1}$ is equivalent to μ, i.e.,

$$\mu(A) = 0 \iff \mu(\sigma^{-1}(A)) = 0, \quad \forall A \in \mathcal{B}.$$

Let $End(X, \mathcal{B}, \mu)$ denote the set of all non-singular endomorphisms of $(X, \mathcal{B}, \mu)$.

In this book, we consider only non-singular endomorphisms of standard measure spaces. In general, $\sigma^{-1}(\mathcal{B})$ can be arbitrary sigma-subalgebra of $\mathcal{B}$. We will also assume that $(X, \sigma^{-1}(\mathcal{B}))$ and $(X, \sigma^{-1}(\mathcal{B}), \mu_\sigma)$ are standard Borel measure spaces, respectively, where μ_σ is the restriction of μ to $\sigma^{-1}(\mathcal{B})$.

If $\mu(\sigma^{-1}(A)) = \mu(A)$ for any measurable set A, then σ is called a *measure preserving endomorphism*, and μ is called a σ-*invariant measure*.

In some cases, we will also need the notion of a *forward quasi-invariant measure* μ. This means that, for every μ-measurable set A, the set $\sigma(A)$ is measurable and $\mu(A) = 0 \iff \mu(\sigma(A)) = 0$. For an at most countable-to-one non-singular endomorphism σ, this property is automatically true. On the other hand, it is not hard to construct an endomorphism σ of a measure space $(X, \mathcal{B}, \mu)$ such that σ is not forward quasi-invariant with respect to μ.

It is worth noting that, for standard measure spaces $(X, \mathcal{B}, \mu)$ and non-singular σ, $\sigma(A)$ is measurable when σ satisfies the condition: $\mu(B) = 0 \implies \mu(\sigma(B)) = 0$ for any Borel set B.

Lemma 2.2 *Let σ be a surjective endomorphism of a standard Borel space $(X, \mathcal{B})$. Then $M(X)$ always contains a σ-quasi-invariant measure μ.*

Proof Every endomorphism σ generates a countable Borel equivalence relation $E(\sigma)$ whose classes are the orbits of σ (see Sect. 7.2 for more details). Quasi-invariant measures for σ coincide with quasi-invariant measures for $E(\sigma)$. Then we can use [DJK94, Proposition 3.1] where the existence of $E(\sigma)$-quasi-invariant measures was proved.   □

We will keep the following notation for a surjective endomorphism σ of $\mathcal{F}(X, \mathcal{B})$:

$$\mathcal{Q}_- = \{\mu \in M(X) : \mu \circ \sigma^{-1} \sim \mu\},$$

$$\mathcal{Q}_+ = \{\mu \in M(X) : \mu \circ \sigma \sim \mu\}.$$

The latter should be understood as follows: if $A \in \mathcal{B}$ and $\sigma(A) \in \mathcal{B}$, then $\mu(A) = 0$ if and only if $\mu(\sigma A) = 0$. This remark is used in all cases when we work with the measure $\mu \circ \sigma$.

It is known that there are Borel endomorphisms σ of $(X, \mathcal{B})$ for which there exists no *finite* σ-invariant measure, see e.g. [DJK94].

Remark 2.3 Quasi-invariance of μ with respect to an endomorphism σ of $(X, \mathcal{B}, \mu)$ (backward and forward) allows us to define the notion of Radon-Nikodym derivatives of measures $\lambda \circ \sigma^{-1}$ and $\lambda \circ \sigma$ with respect to λ:

$$\theta_\lambda(x) = \frac{d\lambda \circ \sigma^{-1}}{d\lambda}(x)$$

and

$$\omega_\lambda(x) = \frac{d\lambda \circ \sigma}{d\lambda}(x).$$

In other words, for any function $f \in L^1(\lambda)$, one has

$$\int_X f \circ \sigma \, d\lambda = \int_X f\theta_\lambda \, d\lambda$$

and

$$\int_X (f \circ \sigma) \, \omega_\lambda \, d\lambda = \int_X f \, d\lambda.$$

To justify these relations, we observe that λ and $\lambda \circ \sigma$ are well defined measures when they are considered on the subalgebra $\sigma^{-1}(\mathcal{B})$. When σ is forward quasi-invariant with respect to λ, we can uniquely define the $\sigma^{-1}(\mathcal{B})$-measurable function $\omega_\lambda(x)$. Since $\theta_\lambda \circ \sigma$ is also $\sigma^{-1}(\mathcal{B})$-measurable, then, by uniqueness of the Radon-Nikodym derivative, we obtain that

$$\omega_\lambda(x) = \frac{1}{\theta_\lambda}(\sigma x).$$

The following fact is obvious.

Lemma 2.4 *Suppose that $\sigma \in End(X, \mathcal{B}, \mu)$ and ν is a measure equivalent to μ, i.e., there exists a measurable function ξ such that $d\nu(x) = \xi(x)d\mu(x)$. Then σ is also non-singular with respect to ν, and θ_ν is cohomologous to θ_μ, i.e., $\theta_\nu(x) = \xi(\sigma x)\theta_\mu(x)\xi(x)^{-1}$.*

Here we define the most important dynamical properties of endomorphisms.

Definition 2.5 Let $\sigma \in End(X, \mathcal{B}, \mu)$.

(i) The endomorphism σ is called *conservative* if for any set A of positive measure there exists $n > 0$ such that $\mu(\sigma^n(A) \cap A) > 0$.
(ii) The endomorphism σ is called *ergodic* if whenever A is σ-invariant, i.e. $\sigma^{-1}(A) = A$, then either A or $X \setminus A$ is of measure zero.

(iii) For $\sigma \in End(X, \mathcal{B}, \mu)$, one associates the sequence of subalgebras generated by σ:

$$\mathcal{B} \supset \sigma^{-1}(\mathcal{B}) \cdots \supset \sigma^{-i}(\mathcal{B}) \supset \cdots$$

Then $\sigma \in End(X, \mathcal{B}, \mu)$ is called *exact* if

$$\mathcal{B}_\infty := \bigcap_{k \in \mathbb{N}} \sigma^{-k}(\mathcal{B}) = \{\emptyset, X\} \quad \text{mod } 0.$$

Clearly, every exact endomorphism is ergodic.

The nested (filtered) family of sigma-algebras from Definition 2.5 is a recurrent theme in symbolic dynamics, in multiresolution analysis, and in ergodic theory;– for details, see, for instance, [Kak48], [Roh61], [Rue89], [CFS82], [Jor01], [Jor04], [Haw94]. A main theme in our work is to point out that this basic filtered system has three incarnations in our analysis, each important in a systematic study of transfer operators.

In more detail: The starting point for our study of infinite-dimensional analysis of transfer operators is a fixed system $(X, \mathcal{B}, R, \sigma)$ as specified above, i.e., a fixed transfer operator R, subject to the pull-out property for σ, as in Definition 1.1. The three incarnations we have in mind of the scale of sigma-algebras from Definition 2.5 are: (1) measure-theoretic (Chaps. 3 and 4), (2) geometric/symbolic (Chaps. 3, 10, and 12), and (3) operator theoretic (Chaps. 5, 7, and 8). In each of these settings, we show that when $(X, \mathcal{B}, R, \sigma)$ is given, then the system from Definition 2.5 induces corresponding scales of measures, of certain closed subspaces in a suitable universal Hilbert space, and in geometric systems of self-similar scales; referring to (1)–(3), respectively. The details and the applications of these three correspondences will be presented systematically in the respective chapters (below), inside the body of the book.

2.3 Measurable Partition and Subalgebras

We give here a short overview of the theory of measurable partitions, developed earlier in a series of papers by Rohlin (see his pioneering article [Roh49b] and the book [CFS82] for further references). Later on, the ideas and methods of this theory were used in many papers. We refer to the works [VF85, Ver94, Ver01] where the orbit theory of dynamical systems was studied in the framework of sequences of measurable partitions.

Let $\xi = \{C_\alpha : \alpha \in I\}$ be a *partition* of a standard probability measure space $(X, \mathcal{B}, \mu)$ into measurable sets. We will focus on the most interesting case when all sets C_α and the index set I are uncountable (though some endomorphisms, arising in the examples considered below, have finite preimages).

Let A be a Borel set. One says that a set $A = \bigcup_{\alpha \in I'} C_\alpha$ is a ξ-*set* where I' is any subset of I. Let $\mathcal{B}(\xi)$ be the sigma-algebra of ξ-sets. Clearly, $\mathcal{B}(\xi) \subset \mathcal{B}$.

By definition, a partition ξ is called *measurable* if $\mathcal{B}(\xi)$ contains a countable subset (D_i) of ξ-sets such that it separates any two elements C, C' of ξ: there exists $i \geq 1$ such that either $C \subset D_i$ and $C' \subset X \setminus D_i$ or $C' \subset D_i$ and $C \subset X \setminus D_i$.

Any partition ξ defines the *quotient space* X/ξ whose points are elements of ξ. Let π be the natural projection from X to X/ξ, i.e., $\pi(x) = C_x$. To define a measure space $(X/\xi, \mathcal{B}/\xi), \mu_\xi)$, we say that $E \in \mathcal{B}/\xi$, if and only if the ξ-set $\pi^{-1}(E)$ is in $\mathcal{B}$, and the probability measure μ_ξ is defined on $\pi(\mathcal{B}(\xi))$ by setting $\mu_\xi = \mu \circ \pi^{-1}$.

It can be proved that ξ *is measurable if and only if* $(X/\xi, \pi(\mathcal{B}(\xi)), \mu_\xi)$ *is a standard measure space*, see [Roh49b].

More generally, suppose $(X, \mathcal{B}, \mu)$ and $(Y, \mathcal{C}, \nu)$ are two standard measure spaces. Let $\varphi : X \to Y$ be a measurable map. Then the partition $\zeta := \{\varphi^{-1}(y) : y \in Y\}$ is obviously measurable. In particular, φ can be a surjective non-singular endomorphism of $(X, \mathcal{B}, \mu)$. In this case, we see that the partition $\zeta(\varphi) := \{\varphi^{-1}(x) : x \in X\}$ has the following properties

$$X/\zeta(\varphi) = X, \quad \mathcal{B}(\varphi) = \varphi^{-1}(\mathcal{B}), \quad \mu_\varphi = \mu|_{\varphi^{-1}(\mathcal{B})} \tag{2.1}$$

Hence, the partition $\zeta(\varphi)$ is indexed by points of the space X, that is the quotient space X/ζ is identified with X.

Let $Orb_\varphi(x) := \{y \in X : \varphi^m(y) = \varphi^n(x) \text{ for some } m, n \in \mathbb{N}_0\}$ be the orbit of φ through $x \in X$. Then, in contrast to the above partition ζ, the partition of X into orbits of φ is not measurable, in general.

We recall here a few facts and definitions about measurable partitions that will be used below. It is said that a partition ζ *refines* ξ (in symbols, $\xi \prec \zeta$) if every element C of ξ is a ζ-set. If ξ_α is a family of measurable partitions, then their product $\bigvee_\alpha \xi_\alpha$ is a measurable partition ξ which is uniquely determined by the conditions: (i) $\xi_\alpha \prec \xi$ for all α, and (ii) if η is a measurable partition such that $\xi_\alpha \prec \eta$, then $\xi \prec \eta$. Similarly, one defines the intersection $\bigwedge_\alpha \xi_\alpha$ of measurable partitions.

It turns out that every partition ζ has a *measurable hull*, that is a measurable partition ξ such that $\xi \prec \zeta$ and ξ is a maximal measurable partition with this property. In order to illustrate this fact, we consider a measurable automorphism T of a measure space $(X, \mathcal{B}, \mu)$ and define the partition $\zeta(T)$ of X into orbits of T, $\zeta(T)(x) = \{T^i x : i \in \mathbb{Z}\}$. In general, $\zeta(T)$ is not measurable. There exists a measurable partition ξ, the measurable hull of $\zeta(T)$, which is known as the partition of X into *ergodic components* of T. If T is ergodic, then ξ is the trivial partition.

Lemma 2.6 *There is a one-to-one correspondence between the set of measurable partitions of a standard measure space* $(X, \mathcal{B}, \mu)$ *and the set of complete sigma-subalgebras* $\mathcal{A}$ *of* $\mathcal{B}$. *This correspondence is defined by assigning to each partition*

ξ *the sigma-algebra $\mathcal{B}(\xi)$ of ξ-sets. Moreover,*

$$\mathcal{A}(\bigwedge_\alpha \xi_\alpha) = \bigcap_\alpha \mathcal{A}(\xi_\alpha), \qquad \mathcal{A}(\bigvee_\alpha \xi_\alpha) = \bigvee_\alpha \mathcal{A}(\xi_\alpha)$$

where the latter is the minimal sigma-subalgebra that contains all $\mathcal{A}(\xi_\alpha)$.

We need the following classical result due to Rokhlin [Roh49b] about the disintegration of probability measures.

Definition 2.7 For a standard probability measure space $(X, \mathcal{B}, \mu)$ and a measurable partition ξ of X, we say that a collection of measures $(\mu_C)_{C \in X/\xi}$ is a *system of conditional measures* with respect to $((X, \mathcal{B}, \mu), \xi)$ if

  (i) for each $C \in X/\xi$, μ_C is a measure on the sigma-algebra $\mathcal{B}_C := \mathcal{B} \cap C$ such that $(C, \mathcal{B}_C, \mu_C)$ is a standard probability measure space;

 (ii) for any $B \in \mathcal{B}$, the function $C \mapsto \mu_C(B \cap C)$ is μ_ξ-measurable;

(iii) for any $B \in \mathcal{B}$,

$$\mu(B) = \int_{X/\xi} \mu_C(B \cap C) \, d\mu_\xi(C). \tag{2.2}$$

Theorem 2.8 ([Roh49b]) *For any measurable partition ξ of a standard probability measure space $(X, \mathcal{B}, \mu)$, there exists a unique system of conditional measures (μ_C). Conversely, if $(\mu_C)_{C \in X/\xi}$ is a system of conditional measures with respect to $((X, \mathcal{B}, \mu), \xi)$, then ξ is a measurable partition.*

We notice that relation (2.2) can be written as follows: for any $f \in L^1(X, \mathcal{B}, \mu)$,

$$\int_X f(x) \, d\mu(x) = \int_{X/\xi} \left(\int_C f_C(y) \, d\mu_C(y) \right) d\mu_\xi(C) \tag{2.3}$$

where $f_C = f|_C$.

We can apply this theorem to the case of an onto endomorphism $\varphi \in End(X, \mathcal{B}, \mu)$ as described above. Let $\zeta(\varphi)$ be the measurable partition of $(X, \mathcal{B}, \mu)$ into preimages $\varphi^{-1}(x)$ of points $x \in X$ (see (2.1)). Let (μ_C) be the system of conditional measures defined by $\zeta(\varphi)$. Then relation (2.3) has the form

$$\int_X f(x) \, d\mu(x) = \int_X \left(\int_C f_C(y) \, d\mu_C(y) \right) d\mu_\varphi(C) \tag{2.4}$$

In most important cases, the disintegration of a measure is applied to probability (finite) measures. The problem of measure disintegration is discussed in many books and articles. We refer here to [Bog07], [CFS82], [Fab87, Fab00], [Kec95]. The case of an infinite sigma-finite measure was considered by several authors. We refer here to [Sim12]. The result is formulated in a slightly more general terms, in comparison with probability measures.

Let $(X, \mathcal{B}, \mu)$ and $(Y, \mathcal{A}, \nu)$ be standard measure spaces with sigma-finite measures, and suppose that $\pi : X \to Y$ is a measurable map. By definition, a *system of conditional measures* is a collection of measures $(\nu_y)_{y \in Y}$ such that

1. ν_y is a measure on the standard measure space $(\pi^{-1}(y), \mathcal{B} \cap \pi^{-1}(y))$, $y \in Y$;
2. for every $B \in \mathcal{B}$,

$$\mu(B) = \int_Y \nu_y(B) \, d\nu(y).$$

Theorem 2.9 ([Sim12])   *Let $(X, \mathcal{B}, \mu)$ and $(Y, \mathcal{A}, \nu)$ be as above. Suppose that $\widehat{\mu} = \mu \circ \pi^{-1} \ll \nu$. Then there exists a unique system of conditional measures $(\nu_y)_{y \in Y}$ for μ. For ν-a.e., ν_y is a sigma-finite measure.*

The structure of countable-to-one endomorphisms is described in the following result.

Theorem 2.10 ([Roh49b])   *For a countable-to-one endomorphism σ of $(X, \mathcal{B}, \mu)$, there exists a partition $\zeta = (A_1, A_2, \ldots)$ of X into at most countably many elements such that*

(i) $\mu(A_i) > 0$ for all i;
(ii) $\sigma_i := \sigma|_{A_i}$ is one-to-one and A_i is of maximal measure in $X \setminus \bigcup_{j<i} A_j$ with this property. In particular, σ_1 is one-to-one and onto X.

Clearly, the Rohlin partition ζ is finite if σ is bounded-to-one. Let τ_i be a one-to-one Borel map with domain $\sigma_i(A_i)$ such that $\sigma \circ \tau_i = \mathrm{id}$. Then the collection of maps τ_i's represents the inverse branches of σ. They are used in explicit constructions of positive operators related to iterated function systems. This type of endomorphisms arises also as shifts on stationary Bratteli diagrams, see [BK16], [BJ15].

2.4   Solenoids and Applications

To finish this chapter we recall the construction of natural extension of endomorphisms (or a solenoid in other terms). This solenoid construction outlined here was considered in a variety of special cases. In one form or the other, they arise in such earlier papers dealing with applications to IFS theory, see e.g., [AJL16, BFMP09, BHS12, BMPR12, DJ07, Hut96, Jor01, Jor04, JT17, Mau95].

Let σ be an endomorphism of a standard Borel space $(X, \mathcal{B})$. We associate to $((X, \mathcal{B}), \sigma)$ a *solenoid* $Sol_\sigma(X)$ as follows. By definition,

$$Sol_\sigma(X) := \{y = (x_i) \in \prod_{i=0}^{\infty} X : \sigma(x_{i+1}) = x_i, \ i \in \mathbb{N}_0\}.$$

Since $Sol_\sigma(X)$ is a Borel subset of $\prod_{i=0}^{\infty}(X, \mathcal{B})$, any solenoid is a standard Borel space in its turn. If X is a compact space, then $Sol_\sigma(X)$ is also a compact subset.

Furthermore, $Sol_\sigma(X)$ is an invariant subset of $\prod_{i=0}^\infty X$ with respect to the shift $\sigma_0(x_i) = (\sigma x_i)$. We use the notation π_i, $i \in \mathbb{N}_0$, for the projection from $Sol_\sigma(X)$ onto X, $\pi_i((x_i)) = x_i$.

Lemma 2.11 *Let λ be a Borel measure on $(X, \mathcal{B})$. In the above notation, the partition of $Sol_\sigma(X)$ into the fibers $\{\pi_0^{-1}(x) : x \in X\}$ is measurable.*

Starting with a transfer operator system (X, B, σ, R), there is a general procedure for extending to an invertible dynamical system, now realized on an associated solenoid; see the outline here in Lemma 2.12. As documented in the literature (see, for example, [BJ97, BJ02, BJMP05, BMPR12, DJ06, DR07, Dut02, FGKP16, Jor01, Jor04, JT15, JT17]), there are many applications of this construction: (1) the given endomorphism σ lifts in a canonical fashion to an automorphism on the solenoid; (2) under suitable assumption, the given transfer operator system (X, B, σ, R) then admits a realization by unitary operators, again realized on suitable L^2 spaces and realized on the solenoid; and (3) the construction in (2) includes families of generalized wavelets. These wavelet families in turn include as special cases more traditional multi-resolution wavelet constructions considered earlier in the standard Hilbert space $L^2(R^d)$. Under suitable restrictions, in fact, $L^2(R^d)$ embeds naturally in an L^2 space on the solenoid. We shall refer to the cited literature for details regarding (1)–(3), but see also [AJL16].

For the solenoid $Sol_\sigma(X)$, we define a Borel map $\widetilde{\sigma}$ of the solenoid by setting

$$\widetilde{\sigma}(x_0, x_1, x_2, \ldots) = (\sigma(x_0), x_0, x_1, \ldots) \tag{2.5}$$

Lemma 2.12 *The transformation $\widetilde{\sigma} : Sol_\sigma(X) \to Sol_\sigma(X)$ is a one-to-one and onto map, i.e. $\widetilde{\sigma}$ is a Borel automorphism of the solenoid.*

Proof To see this, we set

$$\widetilde{\sigma}^{-1}(x_0, x_1, x_2, \ldots) = (x_1, x_2, \ldots).$$

Then, the relation $\widetilde{\sigma}^{-1}\widetilde{\sigma} = $ id is obvious. On the other hand, for any $(x_0, x_1, x_2, \ldots) \in Sol_\sigma(X)$, we have

$$\widetilde{\sigma}\widetilde{\sigma}^{-1}(x_0, x_1, x_2, \ldots) = \widetilde{\sigma}(x_1, x_2, \ldots) = \widetilde{\sigma}(y_0, y_1, \ldots)$$

$$= (\sigma(y_0), y_0, y_1, \ldots) = (\sigma(x_1), x_1, x_2, \ldots) = (x_0, x_1, x_2, \ldots)$$

where $y_i = x_{i+1}$, $i \geq 0$.   □

We will use this construction below.

Remark 2.13 It is worth noting that, based on the definition of a transfer operator built by a system of conditional measures, see Example 1.6, we can immediately extend the main results of [DJ07, Theorems 3.1, 3.4] to the case of an arbitrary surjective endomorphism σ.

Chapter 3
Positive, and Transfer, Operators on Measurable Spaces: General Properties

Abstract The notions of positive operators and transfer operators are central objects in this book. We will discuss various properties of these operators and their specific realization in the subsequent chapters. Here we first focus on the most general properties and basic definitions related to these operators.

Keywords Positive operators · Transfer operators · Harmonic functions · Koopman operator · Coboundary

While the setting for a study of transfer operators, and more general positive linear operators, is that of a set Y, and a fixed sigma-algebra $\mathcal{A}$, in order to get explicit characterizations, it is useful to restrict attention to *standard Borel spaces*; so the case when $(Y, \mathcal{A})$ is now a pair $(X, \mathcal{B})$ given to be isomorphic to some separable complete metric space (a Polish space) with associated Borel sigma-algebra $\mathcal{B}$; or $(X, \mathcal{B})$ is isomorphic to some uncountable Borel subset of some separable complete metric space with the induced Borel sigma-algebra. Generally we allow for the possibility that X is non-compact.

By a transfer operator in $(X, \mathcal{B})$ we mean a pair (R, σ) satisfying the conditions in Definition 3.1 (i), and (3.1). The starting point in the present chapter is a fixed pair (R, σ) on $(X, \mathcal{B})$, defining a transfer operator; and we begin with a systematic study of various sets of measures on $(X, \mathcal{B})$ which allow us to derive spectral theoretic information for the transfer operator (R, σ) under consideration. For this purpose, we also make precise a notion of *isomorphisms* of pairs of transfer operators (R, σ); see Definition 3.8. Our study of *measure classes* associated to a fixed (R, σ) will be undertaken in the two chapters to follow.

3.1   Transfer Operators on Borel Functions

We will consider positive and transfer operators acting in some natural spaces $\mathcal{F}$ of real-valued functions. Examples of such spaces are: $\mathcal{F}(X, \mathcal{B})$, $L^p(X, \mathcal{B}, \mu)$ $(1 \leq p \leq \infty)$, $C(X)$ (if X is a Hausdorff topological space), $C_c(X)$ (if X is locally compact), etc. If the domain of R is not specified that it will always be assumed that R is defined on $\mathcal{F}(X, \mathcal{B})$. In all those cases, we write $\mathcal{F}_+$ for the cone of non-negative functions. Hence, we can define a *positive operator* P as a linear operator that preserves the cone of non-negative functions: $f \in \mathcal{F}_+ \implies P(f) \in \mathcal{F}_+$. If a Borel measure μ is given on $(X, \mathcal{B})$, then we can consider non-negative elements of the space $L^p(X, \mathcal{B}, \mu)$ and define a positive operator P on $L^p(\mu)$ similarly.

Let $(X, \mathcal{B})$ be a standard Borel space, and let σ be a surjective Borel endomorphism of $(X, \mathcal{B})$. A function f is called $\sigma^{-1}(\mathcal{B})$-measurable if $f \in \mathcal{F}(X, \sigma^{-1}(\mathcal{B}))$. For any function $f \in \mathcal{F}(X)$, the function $f \circ \sigma$ is constant on every element of the partition $\xi = \{\sigma^{-1}(x) : x \in X\}$, and therefore $f \circ \sigma$ is measurable with respect to $\sigma^{-1}(\mathcal{B})$. Thus, it can be easily seen that a Borel function g is $\sigma^{-1}(\mathcal{B})$-measurable if and only if there exists a Borel function G such that $g = G \circ \sigma$. In this settings, the operator $U : \mathcal{F}(X, \mathcal{B}) \to \mathcal{F}(X, \sigma^{-1}(\mathcal{B})) : f \mapsto f \circ \sigma$ is positive and called the *composition operator* or *Koopman operator*. In the framework of ergodic theory this operator U being considered on the spaces $L^1(\mu)$ or $L^2(\mu)$ is known by the name of *Koopman operator*, see e.g., [Rue78, CFS82]. We also refer to [Kar59] as one of the pioneering papers on positive operators.

The set of positive operators contains an important class of operators called *transfer operators*. We find it useful to expand the definition of a transfer operator from Chap. 1, giving more details now.

Definition 3.1

(1) Let $\sigma : X \to X$ be a surjective endomorphism of a standard Borel space $(X, \mathcal{B})$. We say that R is a *transfer operator* if $R : \mathcal{F}(X) \to \mathcal{F}(X)$ is a linear operator satisfying the properties:

   (i) $f \geq 0 \implies R(f) \geq 0$ (i.e., R is positive);
   (ii) for any Borel functions $f, g \in \mathcal{F}(X)$,

$$R((f \circ \sigma)g) = f R(g). \qquad (3.1)$$

(2) For a non-singular endomorphism σ on $(X, \mathcal{B}, \mu)$, we define similarly a transfer operator acting in the space $L^p(X, \mathcal{B}, \mu)$, $1 \leq p \leq \infty$.
(3) If $R(\mathbf{1})(x) > 0$ for all $x \in X$, then we say that R is a *strict* transfer operator (here and below the expression $R(\mathbf{1})$ means the image of the constant function $\mathbf{1} : x \mapsto 1$).
(4) If $R(\mathbf{1}) = \mathbf{1}$, then the transfer operator R is called *normalized*.
(5) If h is a non-negative function such that $Rh = h$, then h is called a *harmonic function*.

The condition (3.1) implies that $R(f \circ \sigma) = f R(\mathbf{1})$. In particular, if R is normalized, then $R(f \circ \sigma) = f$.

We use also the notation (R, σ) for a transfer operator R to emphasize that these two objects are closely related according to the "pull-out property" given in (3.1). Moreover, this point of view is useful for the problem of classification of transfer operators (see the corresponding definitions below in this section). It is worth remarking that the set $\mathcal{R}(\sigma)$ of transfer operators R defined by the same endomorphism σ can be vast.

Remark 3.2

(1) Since we work with standard Borel and measure spaces, the transfer operators do not depend on underlying space, in general. This means that if $(X, \mathcal{B})$ and $(Y, \mathcal{A})$ are standard Borel spaces and $\psi : (X, \mathcal{B}) \to (Y, \mathcal{A})$ is a Borel map implementing the Borel isomorphism of these spaces, then, for every transfer operator (R, σ) acting in $\mathcal{F}(X, \mathcal{B})$, there exists an isomorphic transfer operator (R', σ') acting on the space $\mathcal{F}(Y, \mathcal{A})$. We discuss the notion of isomorphism of transfer operators below in this section.

(2) When we discuss properties of a transfer operator R, we will mostly work with non-negative Borel (or measurable) functions. The point is that if a transfer operator R is defined on the cone of positive functions $\mathcal{F}(X)_+$, then R is naturally extended to $\mathcal{F}(X)$ by linearity. The same approach is used in all statements related to integration with respect to a measure λ.

The point of view on transfer operators as pairs (R, σ) allows us to introduce a semigroup structure on such pairs.

Let $\sigma \in End(X, \mathcal{B})$, and let

$$\mathcal{R}(\sigma) := \{(R, \sigma) : R \text{ is a transfer operator w.r.t. } \sigma\}.$$

Denote

$$\mathcal{R}(X, \mathcal{B}) := \bigcup_{\sigma \in End(X, \mathcal{B})} \mathcal{R}(\sigma).$$

Lemma 3.3

(1) The set $\mathcal{R}(X, \mathcal{B})$ is a semigroup with identity with respect to the product

$$(R_1 R_2, \sigma_1 \sigma_2) = (R_1, \sigma_1)(R_2, \sigma_2).$$

(Here the notation $R_1 R_2$ and $\sigma_1 \sigma_2$ means the composition of mappings.)
(2) The set $\mathcal{R}(\sigma)$ is a vector space for each fixed $\sigma \in End(X, \mathcal{B})$.

Proof Let (R_1, σ_1) and (R_2, σ_2) be two transfer operators, where $R_i : \mathcal{F}(X, \mathcal{B}) \to \mathcal{F}(X, \mathcal{B})$ and σ_i is an onto endomorphism of $(X, \mathcal{B})$, $i = 1, 2$. We need to check that $(R_1 R_2, \sigma_1 \sigma_2)$ is a well defined transfer operator in $\mathcal{F}(X, \mathcal{B})$. Since the range of any transfer operator (R, σ) is $\mathcal{F}(X, \mathcal{B})$ (see Lemma 3.11), the composition $R_1 R_2$ is defined. It remains to check that $(R_1 R_2, \sigma_1 \sigma_2)$ satisfies Definition 3.1. The positivity is obvious and

$$R_1 R_2[f(\sigma_1 \sigma_2(x))g(x)] = R_1[f(\sigma_1(x))R_2(g(x))]$$
$$= f(x)R_1 R_2(g)(x).$$

The second claim is clear because, for $a, b \in \mathbb{R}$,

$$(aR_1 + bR_2)[(f \circ \sigma)g] = af R_1(g) + bf R_2(g)$$
$$= f(aR_1 + bR_2)(g).$$

□

The dynamical properties of endomorphisms σ such as ergodicity, mixing, etc can be described in terms of transfer operators, see [LM94, DZ09]. We mention here several simple observations to motivate our future study.

Remark 3.4

(1) Suppose that (R, σ) is a transfer operator acting on the space $(X, \mathcal{B}, \mu)$. If σ is not an ergodic endomorphism of $(X, \mathcal{B}, \mu)$, then, for any σ-invariant set A of positive measure $(\sigma^{-1}(A) = A \mod 0)$, we can define the restriction of R on $(A, \mathcal{B}|_A)$. For this, we set

$$R_A(f) = R(\chi_A f), \qquad f \in \mathcal{F}(X, \mathcal{B}).$$

(2) Suppose that σ is periodic on $(X, \mathcal{B})$ of period p, i.e., $\sigma^p(x) = x$ for all x. If R is a transfer operator from $\mathcal{R}(\sigma)$, then R is also periodic. Indeed, R^p is a transfer operator corresponding σ^p. Hence, for any functions $f, g \in \mathcal{F}(X, \mathcal{B})$, it satisfies the relation $R^p(f)g = f R^p(g)$ which means that R^p is the identity operator.

Lemma 3.5 *The operator $R_A : \mathcal{F}(A, \mathcal{B}|_A) \to \mathcal{F}(A, \mathcal{B}|_A)$ is a transfer operators corresponding to $\sigma_A = \sigma : A \to A$.*

Proof We need to check that (R_A, σ_A) satisfies the Definition 3.1:

$$R_A((f \circ \sigma)g) = R(\chi_A(f \circ \sigma)g)$$

$$= R((\chi_A)^2(f \circ \sigma)g)$$

$$= R((\chi_A \circ \sigma)(f \circ \sigma)\chi_A g)$$

$$= \chi_A f R(\chi_A g)$$

$$= \chi_A f R_A(g).$$

We used here the relation $\chi_A = \chi_{\sigma^{-1}(A)} = \chi_A \circ \sigma$.   □

The statements, formulated in Lemma 3.5 and Remark 3.4, mean that the classes of all ergodic endomorphisms of a measure space, and the aperiodic endomorphisms for Borel spaces, play the central role. Thus, we can avoid some trivialities by considering only ergodic and/or aperiodic endomorphisms σ.

3.2 Classification

The problem of classification of transfer operators has many aspects and depends on the choice of equivalence relations on the set of all transfer operators. We consider only the definition of isomorphic transfer operators (R, σ). For motivation, we begin with the following example.

Example 3.6 Suppose that σ and σ' are two surjective endomorphisms of $(X, \mathcal{B})$ such that $\sigma'\tau(x) = \tau\sigma(x)$ for some one-to-one Borel map τ and all $x \in X$. We define the operator $S := S_\tau$ acting on the set of Borel functions $f \in \mathcal{F}(X, \mathcal{B})$ by the formula

$$(Sf)(x) := f(\tau x).$$

Lemma 3.7 *Let (R, σ) be a transfer operator in $\mathcal{F}(X, \mathcal{B})$. Then $R' = S^{-1}RS$ is the transfer operator corresponding to the endomorphism σ'.*

Proof This observation follows from the facts that R' is positive and (R', σ') satisfies the relation:

$$S^{-1}RS[f(\sigma'x)g(x)] = S^{-1}R[f(\sigma'\tau x)g(\tau x)]$$

$$= S^{-1}R[f(\tau(\sigma x))g(\tau x)]$$

$$= S^{-1}[f \circ \tau(x)(Rg)(\tau x)]$$

$$= f(x)(S^{-1}RSg)(x).$$

$\square$

The next definition is a generalization of the above example.

Definition 3.8 Let σ_i be an onto endomorphism of a standard Borel space $(X_i, \mathcal{B}_i)$, where $i = 1, 2$. Suppose that (R_1, σ_1) and (R_2, σ_2) are transfer operators acting on Borel functions defined on $(X_1, \mathcal{B}_1)$ and $(X_2, \mathcal{B}_2)$, respectively. We say that (R_1, σ_1) and (R_2, σ_2) are *isomorphic* if there exists a Borel isomorphism $T : (X_1, \mathcal{B}_1) \rightarrow (X_2, \mathcal{B}_2)$ such that

$$T\sigma_1 = \sigma_2 T \quad \text{and} \quad T_*R_2 = R_1 T_*,$$

where T_* is the induced map $\mathcal{F}(X_2) \rightarrow \mathcal{F}(X_1)$:

$$(T_*f)(x_1) = f(Tx_1), \quad \forall f \in \mathcal{F}(X_2), \ x_1 \in X_1.$$

In order to justify this definition, we need to show that $T R_2 T^{-1}$ is a transfer operator corresponding to σ_1. It is obvious that $T R_2 T^{-1}$ is a positive operator. To verify the pull-out property (1.1), we calculate, for $g, h \in \mathcal{F}(X_1)$,

$$
\begin{aligned}
T R_2 T^{-1}[(g \circ \sigma_1) h] &= T R_2[g(\sigma_1 T^{-1} x) h(T^{-1} x)] \\
&= T_* R_2[(g \circ T_*^{-1}(\sigma_2 x))(h \circ T^{-1})(x)] \\
&= T_*[(g \circ T^{-1}(x)) R_2(h \circ T^{-1})(x)] \\
&= g(x)(T_* R_2 T_*^{-1} h)(x).
\end{aligned}
$$

If $T : X_1 \to X_2$ is not an invertible Borel map, then this definition gives the notion of a *factor* map between two transfer operators.

Example 3.9 In this example, we illustrate the definition of the *isomorphism* for the transfer operators defined by the formula

$$
(R_i f)(x) := \sum_{\sigma_i y = x} q_i(y) f(y), i = 1, 2. \tag{3.2}
$$

Here σ_i is a finite-to-one onto endomorphism of X_i. *Under what conditions on q_1, q_2 are the transfer operators (R_1, σ_1) and (R_2, σ_2) isomorphic?* Let $T : X_1 \to X_2$ be as in Definition 3.8. If one rewrites the relation $(T_* R_2 f)(x) = (R_1 T_* f)(x)$ with $f \in \mathcal{F}(X_2)$, then it transforms to the identity

$$
\sum_{y: \sigma_2 y = T x} q_2(y) f(y) = \sum_{z: \sigma_1 z = x} q_1(z) f(T z)
$$

which holds for any $x \in X_1$ and any Borel function $f \in \mathcal{F}(X_2)$.

Lemma 3.10 *Let (R_1, σ_1) and (R_2, σ_2) be defined by (3.2). If they are isomorphic via a transformation T, then*

$$
\sum_{a \in C_x} q_2(a) = \sum_{a \in C_x} (T_*^{-1} q_1)(a) \tag{3.3}
$$

where $C_x = \{a \in X_2 : T^{-1} \sigma_2 a = x\}$.

Proof Take $f = \delta_a$ where a is a point from X_2. Then

$$
(T_* R_2 \delta_a)(x) = \sum_{a: \sigma_2 a = T x} q_2(a)
$$

and

$$(R_1 T_* \delta_a)(x) = \sum_{a:\sigma_1 T^{-1}a=x} q(T^{-1}a).$$

We notice that $T^{-1}\sigma_2 a = \sigma_1 T^{-1}a$, therefore the relation $(T_* R_2 f)(x) = (R_1 T_* f)(x)$ implies (3.3). $\qquad\square$

3.3 Kernel and Range of Transfer Operators

In the next statements we discuss the *structural properties* of transfer operators.

Lemma 3.11 *Let σ be an onto endomorphism of a standard Borel space $(X, \mathcal{B})$. Suppose $R : \mathcal{F}(X, \mathcal{B}) \to \mathcal{F}(X, \mathcal{B})$ is a strict transfer operator with respect to σ. If $R|_\sigma$ is the restriction of R onto $\mathcal{F}(X, \sigma^{-1}(\mathcal{B}))$, then*

$$R|_\sigma : \mathcal{F}(X, \sigma^{-1}(\mathcal{B})) \to \mathcal{F}(X, \mathcal{B})$$

is a one-to-one and onto map.

Proof Since, for any function $f \in \mathcal{F}(X, \mathcal{B})$,

$$R(f \circ \sigma) = f R(\mathbf{1}) \tag{3.4}$$

and $R(\mathbf{1}) > 0$, we see that $R|_\sigma$ is onto.

Suppose $f, g \in \mathcal{F}(X, \mathcal{B})$ are two distinct Borel functions and set $A = \{x \in X : f(x) \neq g(x)\}$. Then, for $x \in \sigma^{-1}(A)$, we have $(f \circ \sigma)(x) \neq (g \circ \sigma)(x)$. It follows that

$$R(f \circ \sigma) = f R(\mathbf{1}) \neq g R(\mathbf{1}) = R(g \circ \sigma)$$

and the proof is complete. $\qquad\square$

Denote by $\mathcal{S}(X)$ the set of real-valued simple functions on $(X, \mathcal{B})$:

$$\mathcal{S}(X) := \{s : X \to \mathbb{R} : s(x) = \sum_{i \in I} c_i \chi_{E_i}, \; |I| < \infty\}$$

where $\{E_i : i \in I\}$ is any finite partition of X into Borel subsets.

Lemma 3.12 *Suppose σ is an onto endomorphism of a standard Borel space. Let $R : \mathcal{F}(X, \mathcal{B}) \to \mathcal{F}(X, \mathcal{B})$ be a normalized transfer operator. Then R sends the set of $\sigma^{-1}(\mathcal{B})$-measurable simple functions onto the set of simple functions in $\mathcal{F}(X, \mathcal{B})$ (by Lemma 3.11 this map is one-to-one and onto).*

Proof The result follows from the following observation: for any set $A \in \mathcal{B}$,

$$R(\chi_{\sigma^{-1}(A)}) = R(\chi_A \circ \sigma) = \chi_A R(\mathbf{1}) = \chi_A.$$

Hence, this relation is extended to simple functions by linearity. $\square$

Based on the proved results, one can ask whether relation (3.4) determines the pull-out property. The affirmative answer is contained in the following lemma.

Lemma 3.13 *If* $(X, \mathcal{B})$ *and* σ *are as above, then*

$$R(f \circ \sigma) = f R(\mathbf{1}) \iff R((f \circ \sigma)g) = f R(g), \quad \forall g \in \mathcal{F}(X, \sigma^{-1}(\mathcal{B})).$$

Proof We need to show only that ($\Longrightarrow$) holds. Indeed, if $g \in \mathcal{F}(X, \sigma^{-1}(\mathcal{B}))$, then there exists $G \in \mathcal{F}(X, \mathcal{B})$ such that $g = G \circ \sigma$. Then

$$R((f \circ \sigma)g) = R((fG) \circ \sigma) = fGR(\mathbf{1}) = f R(G \circ \sigma) = f R(g).$$

$\square$

Consider the kernel of R,

$$Ker(R) := \{f \in \mathcal{F}(X, \mathcal{B}) : R(f) = 0\}.$$

It is clear that the pull-out property implies that

$$f \in Ker(R) \Longrightarrow f(g \circ \sigma) \in Ker(R), \quad \forall g \in \mathcal{F}(X, \sigma^{-1}(\mathcal{B})).$$

The above relation shows that $Ker(R)$ can be viewed as an $\mathcal{F}(X, \sigma^{-1}(\mathcal{B}))$-module.

Theorem 3.14 *Let* (R, σ) *be a normalized transfer operator on* $\mathcal{F}(X, \mathcal{B})$ *where* σ *is an onto endomorphism. For any Borel function* $f \in \mathcal{F}(X, \mathcal{B})$, *there exist uniquely determined functions* $f_0 \in Ker(R)$ *and* $\overline{f} \in \mathcal{F}(X, \sigma^{-1}(\mathcal{B}))$ *such that*

$$f = f_0 + \overline{f}. \tag{3.5}$$

Proof We first show that, for any Borel function $f \notin \mathcal{F}(X, \sigma^{-1}(\mathcal{B}))$, there exists a function $\overline{f} \in \mathcal{F}(X, \sigma^{-1}(\mathcal{B}))$ such that $R(f) = R(\overline{f})$. (If $f \in \mathcal{F}(X, \sigma^{-1}(\mathcal{B}))$, the we take $\overline{f} = f$.) Indeed, take $R(f)$ and set $\overline{f} = R(f) \circ \sigma$. Since $R(\mathbf{1}) = \mathbf{1}$, the function $\overline{f}$ has the desired properties. Set $f_0 = f - \overline{f}$. Then $f_0 \in Ker(R)$ and (3.5) is proved.

It remains to show that this representation is unique. If $f = f = f_0 + \overline{f} = g_0 + \overline{g}$ where $f_0, g_0 \in Ker(R)$, then $R(\overline{g} - \overline{f}) = 0$. Since R is one-to-one on $\mathcal{F}(X, \sigma^{-1}(\mathcal{B}))$, we obtain that $\overline{f} = \overline{g}$ and hence $f_0 = g_0$. $\square$

Corollary 3.15

(1) For a normalized transfer operator (R, σ) as above, let

$$E : \mathcal{F}(X, \mathcal{B}) \to \mathcal{F}(X, \sigma^{-1}(\mathcal{B})) : f \mapsto R(f) \circ \sigma. \tag{3.6}$$

Then the operator E has the following properties: E is positive, $E(\mathcal{F}(X, \mathcal{B})) = \mathcal{F}(X, \sigma^{-1}(\mathcal{B}))$, $E^2 = E$, $E|_{\mathcal{F}(X, \sigma^{-1}(\mathcal{B}))} = id$, $E \circ R = R$, and $R \circ E = R$.
(2) For $Ker(R^n)$ and $\mathcal{F}(X, \sigma^{-n}(\mathcal{B}))$ and any $f \in \mathcal{F}(X, \mathcal{B})$ there exists a decomposition $f = f_0^{(n)} + \overline{f}^{(n)}$ which is similar to (3.5).

Proof Most of the properties formulated in (1) are obvious; we check only that E is an idempotent:

$$E(E(f)) = E(R(f) \circ \sigma) = R[R(f) \circ \sigma] \circ \sigma = R(f) \circ \sigma = E(f).$$

The other relations easily follow from the definition.
   For (2), we notice that

$$Ker(R) \subset Ker(R^2) \subset \cdots \subset Ker(R^n) \subset \cdots$$

and

$$\mathcal{F}(X, \mathcal{B}) \supset \mathcal{F}(X, \sigma^{-1}(\mathcal{B})) \cdots \supset \mathcal{F}(X, \sigma^{-n}(\mathcal{B})) \supset \cdots$$

The proof of the existence of decomposition in (2) is analogous to that in Theorem 3.14.   □

   For an onto endomorphism σ of the space $(X, \mathcal{B})$, we set

$$U_\sigma : \mathcal{F}(X, \mathcal{B}) \to \mathcal{F}(X, \sigma^{-1}(\mathcal{B})) : f(x) \mapsto f(\sigma(x)). \tag{3.7}$$

Corollary 3.16 *For U_σ defined in (3.7), we have*

$$(RU_\sigma)(f) = f, \quad (U_\sigma R)(f) = E(f).$$

If R is not normalized, then the operator RU_σ is the multiplication operator:

$$(RU_\sigma)(f) = R(1)f, \quad f \in \mathcal{F}(X, \mathcal{B}).$$

The restriction of R to $\mathcal{F}(X, \sigma^{-1}(\mathcal{B}))$ is a multiplication operator itself.

   These formulas are obvious. We will use them in the framework of our study of transfer operators in the Hilbert space $L^2(X, \mathcal{B}, \mu)$ where μ is σ-invariant. Then U_σ would be an isometry, and R could be treated as a co-isometry for U_σ. Thus, Corollary 3.16 presents a Borel analogue of the *dual pair* U_σ, R. This observation is the basis for the further study of the isometry U_σ. In particular, in Chap. 7, we discuss the Wold decomposition generated by the sequence of subalgebras $\{\sigma^{-n}(\mathcal{B}) : n \in N_0\}$.

We will also see later that the operator E becomes the conditional expectation when R is considered in the context of $L^p(\mu)$-spaces.

3.4 Multiplicative Properties of Transfer Operators

It turns out that any transfer operator possesses some multiplicative properties when it is restricted to an appropriate subset of Borel functions.

We begin with the following statement proved in [CE77, pp. 166–167] (see also [BJ02, Lemma 5.2.7] (we formulate only a part of the statement here because we need only the fact that A is abelian). Though this result is given in the context of C^*-algebras, it can be easily checked that the proof works in our case. The key ingredient in it is the *Schwarz inequality* for a positive normalized operator R:

$$R(f)^2 \le R(f^2).$$

Lemma 3.17 *Let A be an abelian C^*-algebra with unit 1, and let $E : A \to A$ be a linear map with the properties: (i) E is positive, (ii) $E(1) = 1$, (iii) $E^2 = E$. Then the map*

$$(a, b) \mapsto a \times b := E(ab)$$

is an associate product on the linear space $E(A)$. Moreover, for all $a \in E(A)$ and $b \in A$,

$$E(ab) = E(aE(b)). \tag{3.8}$$

We can apply Lemma 3.17 for the operator E defined in (3.6) (*called a generalized conditional expectation*). It follows from Corollary 3.15 that this operator E satisfies the conditions of the lemma.

Theorem 3.18 *Let R be a normalized transfer operator in $\mathcal{F}(X, \mathcal{B})$. The positive operator $E(f) = R(f) \circ \sigma : \mathcal{F}(X, \mathcal{B}) \to \mathcal{F}(X, \sigma^{-1}(\mathcal{B}))$ satisfies the conditions of Lemma 3.17. For the product $f \times g := E(fg)$, the operator R has the properties:*

$$R(f \times g) = R(f)R(g), \quad f \in \mathcal{F}(X, \sigma^{-1}(\mathcal{B})), g \in \mathcal{F}(X, \mathcal{B}), \tag{3.9}$$

$$R(fg) = R(f)R(g), \quad f, g \in \mathcal{F}(X, \sigma^{-1}(\mathcal{B})).$$

Proof We observe first that $E(f) = R(f) \circ \sigma, f \in \mathcal{F}(X, \mathcal{B})$, is a positive normalized idempotent map onto $\mathcal{F}(X, \sigma^{-1}(\mathcal{B}))$. Thus we can use the conclusion of Lemma 3.17. It follows from (3.8) that, for $f \in \mathcal{F}(X, \sigma^{-1}(\mathcal{B}))$ and $g \in \mathcal{F}(X, \mathcal{B})$,

$$f \times g = E(f E(g)) = R(f(R(g) \circ \sigma)) \circ \sigma = [R(f)R(g)] \circ \sigma$$

Applying R we obtain

$$R(f \times g) = R(f)R(g).$$

Here the relation $R(\mathbf{1}) = \mathbf{1}$ has been repeatedly used.

Since $\mathcal{F}(X, \sigma^{-1}(\mathcal{B})) = \{f \circ \sigma : f \in \mathcal{F}(X, \mathcal{B})\}$, we see that

$$(f \circ \sigma) \times (g \circ \sigma) = E((f \circ \sigma)(g \circ \sigma))$$

$$= R[(f \circ \sigma)(g \circ \sigma)] \circ \sigma$$

$$= (fg) \circ \sigma = (f \circ \sigma)(g \circ \sigma).$$

Now we use the proved relation to conclude that $R(fg) = R(f)R(g)$ holds on the space of Borel functions measurable with respect to $\sigma^{-1}(\mathcal{B})$. $\qquad\square$

Remark 3.19

(1) We emphasize that, though the product $f \times g$ is defined for functions from $\mathcal{F}(X, \mathcal{B})$, the multiplicative property of R is only true when at least one function belongs to $\mathcal{F}(X, \sigma^{-1}(\mathcal{B}))$.
(2) To illustrate Theorem 3.18, we show that (3.9) holds for the transfer operator $R(\sigma)$ (see (1.2)) under the condition $\sum_{y:\sigma(y)=x} W(y) = 1$ for all x. Then

$$R_\sigma((f \circ \sigma)g) = \sum_{y:\sigma(y)=x} W(y)f(\sigma(y))g(y)$$

$$= f(x) \sum_{y:\sigma(y)=x} W(y)g(y)$$

$$= \sum_{y:\sigma(y)=x} W(y)f(\sigma(y)) \sum_{y:\sigma(y)=x} W(y)g(y)$$

$$= R_\sigma(f \circ \sigma)R_\sigma(g).$$

3.5 Harmonic Functions and Coboundaries for Transfer Operators

Harmonic functions will be discussed in the book repeatedly. We mention first a couple of simple facts about characteristic functions.

In what follows, we will work with transfer operators. In this case, the study of harmonic functions is more interesting. We begin with a simple result about σ-invariant functions.

Lemma 3.20

(1) Let σ be a surjective endomorphism of a standard Borel space $(X, \mathcal{B})$. Suppose $s = \sum_i c_i \chi_{A_i}$ is a non-negative simple function such that $\sigma^{-1}(A_i) = A_i$ for all i. Then s is a harmonic function for any normalized transfer operator $(R, \sigma) \in \mathcal{R}(\sigma)$.

(2) Let $G = \{g \in \mathcal{F}(X) : g \circ \sigma = g\}$ be the set of σ-invariant functions. Then G is contained in the space of harmonic function with respect to any normalized transfer operator (R, σ).

Proof

(1) We notice that if $\sigma^{-1}(A) = A$, then $\chi_A(x) = \chi_{\sigma^{-1}(A)}(x) = \chi_A(\sigma x)$. Therefore,

$$R(s) = R(\sum_i c_i \chi_{A_i}) = R(\sum_i c_i \chi_{A_i} \circ \sigma) = s.$$

(2) The statement can be proved similarly to (1).   □

We observe that the proof, given in this lemma, works for any harmonic function.

We have already noticed how important is the property $R(1) = 1$ in the study of transfer operators. In the next lemma, we give simple conditions under which a transfer operator can be normalized.

Lemma 3.21

(1) If (R, σ) is a strict transfer operator acting in the space $\mathcal{F}(X, \mathcal{B})$, then $R_1 = (R1)^{-1} R$ is a normalized transfer operator.

(2) Let (R, σ) be a transfer operator, and let k be a non-negative Borel function. Then the operator

$$R_k(f)(x) = \begin{cases} \dfrac{R(fk)}{k}(x), & \text{if } x \in \{k \neq 0\} \\ 0, & \text{if } x \in \{k = 0\} \end{cases} \tag{3.10}$$

is a well defined transfer operator. Moreover, R_k is a normalized transfer operator if and only if k is a harmonic function for R.

Proof The first statement is trivial. In order to prove the second one, we check that (R_k, σ) satisfies the definition of a transfer operator. Indeed, the positivity of R_k is clear, and the pull-out property follows from the relation

$$R_k((f \circ \sigma)g) = \frac{R((f \circ \sigma)gk)}{k} = f\frac{R(gk)}{k} = fR_k(g).$$

To finish the proof, we observe that the property $R_k(1) = 1$ holds if and only if k is a harmonic function for R. □

Remark 3.22

(1) We note that the operator R_k can be considered as an abstract *Doob transform*. We refer, for instance, to the papers [AU15, AGZ15] for more details.
(2) It is useful to justify the correctness of the definition of R_k in (3.10). For this, we notice that in case when h is a harmonic function for R, then

$$h(x) = 0 \implies R(fh)(x) = 0.$$

Indeed, since R is positive, the *Schwarz inequality* shows that

$$|R(fh)| \leq \sqrt{R(f^2)}\sqrt{R(h^2)} \leq \sqrt{R(f^2)R(h)} = \sqrt{R(f^2)}h,$$

and the result follows.

In fact, one can prove even a stronger result. Write $fh = f\sqrt{h}\sqrt{h}$, and apply the Scwarz inequality

$$|R(fh)| \leq R(f^2h)^{1/2}R(h)^{1/2} \leq R(f^2h)^{1/2}h^{1/2}.$$

Repeating this inequality k times, we obtain

$$|R(fh)| \leq R(f^{2^k}h)^{2^{-k}}h^{2^{-1}+\cdots+2^{-k}}.$$

Remark 3.23 Suppose (R_1, σ_1) and (R_2, σ_2) are isomorphic transfer operators. Let $T : X_1 \to X_2$ be a one-to-one Borel map that implements the isomorphism. This means, in particular, that $T_*R_2 = R_1T_*$ where T_* is the induced map from $\mathcal{F}(X_2)$ to $\mathcal{F}(X_1)$. Then T_* realizes a one- to-one correspondence between the sets of harmonic functions $H(R_2)$ and $H(R_1)$. Indeed, let $h \in H(R_2)$ be a R_2-harmonic function. Then $T_*h = T_*R_2h = R_1T_*h$.

If $Rh = h$ for a transfer operator R, then this fact can be treated as the existence of an eigenfunction h corresponding to the eigenvalue 1. Similarly, we can represent $H(R)$ as the eigenspace corresponding to the eigenvalue 1.

The following observation is based on Corollary 3.16.

Lemma 3.24 *Let (R, σ) be a transfer operator acting in the space of functions $\mathcal{F}(X, \mathcal{B})$. Then $\mathcal{F}(X, \sigma^{-1}(\mathcal{B}))$ is the eigenspace corresponding to the eigenvalue $R(1)$, that is $\mathcal{F}(X, \sigma^{-1}(\mathcal{B})) = \{h \in \mathcal{F}(X) : Rh = R(1)h\}$.*

We say that two transfer operators R_1 and R_2 from $\mathcal{R}(\sigma)$ are *equivalent* ($R_1 \sim R_2$ in symbols) if there exists a non-negative $k \in \mathcal{F}(X, \mathcal{B})$ such that

$$k R_2(f) = R_1(kf), \quad \forall f \in \mathcal{F}(X, \mathcal{B}). \tag{3.11}$$

Let Γ denote the multiplicative group $\mathcal{F}(X, \mathcal{B})_+^0$ of strictly positive Borel functions. For the sake of simplicity, we will assume that the function k from (3.11) is taken from Γ though this restriction can be, in general, omitted.

Proposition 3.25

(1) $R_1 \sim R_2$ is an equivalence relation on the set $\mathcal{R}(\sigma)$ of transfer operators.
(2) $(R_k)_g = (R_g)_k = R_{kg}$ where $k, g \in \mathcal{F}(X, \mathcal{B})_+$.
(3) The map $r(k, R) \mapsto R_k$ defines an action of $\Gamma = \mathcal{F}(X, \mathcal{B})_+^0$ on the set $\mathcal{R}(\sigma)$. The equivalence relation $\sim$ coincides with the partition into orbits of the action r.
(4) The action of Γ on $\mathcal{R}(\sigma)$ is not free. If $\Gamma_0 = \Gamma \cap \{g \in \mathcal{F}(X, \mathcal{B}) : g \circ \sigma = g\}$, then Γ_0 belongs to the stabilizer of every $R \in \mathcal{R}(\sigma)$.

Proof Statement (1) is verified directly. If $R_1(f) = \dfrac{R_2(kf)}{k}$, then, denoting $g = kf$, we get

$$R_2(g) = \frac{R_1(gk^{-1})}{k^{-1}};$$

this proves that $\sim$ is symmetric. Clearly, if $R_1 \sim R_2$ and $R_2 \sim R_3$, then $R_1 \sim R_3$.
    (2) follows from (1).
    To see that (3) holds, we use (2) and compute

$$r(k_1, r(k_2, R)) = r(k_1, R_{k_2}) = (R_{k_2})_{k_1} = R_{k_1 k_2}.$$

Hence, for a fixed $R \in \mathcal{R}(\sigma)$, the set $\{R_k : k \in \mathcal{F}(X, \mathcal{B})_+^0\}$ is the orbit of the action of $\Gamma = \mathcal{F}(X, \mathcal{B})_+^0$ on the set $\mathcal{R}(\sigma)$.
    (4) The action of Γ is not free: $k(I) = I, \forall k \in \Gamma$ where $I(f) = f$ (the identity map). Moreover, if $k \in \Gamma_0$, then $k \circ \sigma = k$ and we have

$$(kR)(f) = R_k(f) = \frac{R((k \circ \sigma)f)}{k} = R(f)$$

where $R \in \mathcal{R}(\sigma)$ is a fixed transfer operator and f is any Borel function.   □

Lemma 3.26 *For $\sigma \in End(X, \mathcal{B})$, let R be a transfer operator from $\mathcal{R}(\sigma)$. If $k \in \Gamma$, then the map $h \mapsto kh$ sends the set $H(R_k)$ onto $H(R)$.*

Proof We need to show that $h \in H(R_k)$ if and only if $hk \in H(R)$. This result follows from the relation:

$$R_k(h) = \frac{R(kh)}{k} = h \iff R(hk) = hk.$$

$\square$

Definition 3.27 Let (R, σ) be a transfer operator. We say that a function $f \in \mathcal{F}(X, \mathcal{B})$ is a σ-coboundary if there exists some $g \in \mathcal{F}(X, \mathcal{B})$ such that $(g \circ \sigma)f = g$.

We say that $f \in \mathcal{F}(X, \mathcal{B})$ is an R-coboundary if there exists $k \in \mathcal{F}(X, \mathcal{B})$ such that $kf = R(k)$. The set of all R-coboundaries is denoted by $Cb(R)$.

From Definition 3.27 we can deduce the following result.

Proposition 3.28 *Let R be a normalized operator from $\mathcal{R}(\sigma)$.*

(1) If f is a σ-coboundary, then $R(f)$ is an R-coboundary.
(2)

$$Cb(R) = Cb(R_k), \qquad \forall k \in \Gamma.$$

Proof For (1), we simply apply R to the equality $(g \circ \sigma)f = g$ and obtain the result. The converse is not true, in general.

Let f be an R-coboundary, i.e., there exists some $g \in \mathcal{F}(X, \mathcal{B})$ such that $gf = R(g)$. Fix $k \in \Gamma$. We claim that $g \in Cb(R_k)$. Indeed, we need to show that there exists h such that $hf = R_k(h)$ or, equivalently,

$$hf = \frac{R(hk)}{k}.$$

The latter means that h must satisfy the equality

$$hkf = R(hk).$$

If we take $h = gk^{-1}$, then we get $gf = R(g)$ which is true. Hence, $Cb(R) \subset Cb(R_k)$. The converse inclusion is proved similarly. $\square$

Let f be a Borel function on $(X, \mathcal{B})$, and let $\sigma : X \to X$ be an onto endomorphism of $(X, \mathcal{B})$. Define the semigroup *cocycle* generated by (f, σ):

$$\alpha_f(x, \sigma^k) := f(\sigma^{k-1}(x))f(\sigma^{k-2}(x)) \cdots f(x), \quad k \in \mathbb{N}.$$

We observe that the following fact holds.

Lemma 3.29 *Suppose that (R, σ) is a transfer operator acting in $\mathcal{F}(X, \mathcal{B})$, and let h be an R-harmonic function. Then*

$$R(\alpha_h(x, \sigma^k)) = (h(x))^k, \quad k \in \mathbb{N}.$$

Proof We calculate

$$R(\alpha_h(x, \sigma)) = R((h \circ \sigma)h)(x) = h(x)R(h)(x) = h^2(x).$$

Then the result follows by induction.   □

Chapter 4
Transfer Operators on Measure Spaces

Abstract Our starting point is a fixed pair (R, σ) on $(X, \mathcal{B})$ making up a transfer operator. In the next two chapters we turn to a systematic study of specific and important *sets of measures* on $(X, \mathcal{B})$ and actions of (R, σ) on these sets of measures. These classes of measures in turn lead to a *structure theory* for our given transfer operator (R, σ). Our corresponding structure results are Theorems 4.14, 5.13, 5.12, 5.9, and 5.20.

Keywords Transfer operators · Banach lattices · Structure theory

4.1 Transfer Operators and Measures

In general, positive and transfer operators on the space $\mathcal{F}(X, \mathcal{B})$ or $L^p(X, \mathcal{B}, \mu)$ are not continuous (see Definition 4.1 for detail). But one can use the notion of *order convergence* to define the order continuity of positive operators. Under this assumption we can define an action of a positive operator on the set $M(X)$ of Borel measures. Order continuity of a positive operator is commonly used, in particular, for positive operators on Banach lattices. The literature devoted to this subject is very extensive; we refer to [AA01] for details and further references. Readers coming from other but related areas, may find the following papers/books useful for background [DJ15, Fed13, ZJ15, JS15, LP15, Mai13, RG16].

In this section, we work with *positive* and *transfer* operators acting on the space of measurable (or integrable) functions on a standard measure space $(X, \mathcal{B}, \mu)$ where μ is a continuous Borel measure. We use the same notation and definitions as in case of Borel functions keeping in mind the mod 0 convention. By P we denote a positive operator acting on an $L^p(\mu)$-space, $1 \leq p \leq \infty$. If $\sigma \in End(X, \mathcal{B}, \mu)$ is a measurable surjective endomorphism, then we define a transfer operator $R = (R, \sigma)$ on $L^p(\mu)$ as in Definition 3.1. We recall that in this case σ is assumed to be a nonsingular onto endomorphism.

© Springer International Publishing AG, part of Springer Nature 2018   39
S. Bezuglyi, P. E. T. Jorgensen, *Transfer Operators, Endomorphisms,*
and Measurable Partitions, Lecture Notes in Mathematics 2217,
https://doi.org/10.1007/978-3-319-92417-5_4

The next definition is formulated in a setting which is suitable for our purposes. We use the language of Banach lattices keeping in mind the functional spaces $\mathcal{F}(X, \mathcal{B})$ and $L^p(\mu)$.

Definition 4.1 Let $\mathcal{E}$ be a Banach lattice and P a positive operator on $\mathcal{E}$. A sequence (f_n) in $\mathcal{E}$ is *order convergent* to $g \in \mathcal{E}$, written $f_n \xrightarrow{o} g$, if there exists a sequence (h_n) in $\mathcal{E}_+$ such that $h_n \downarrow 0$ and $|f_n - g| \leq h_n$ for all $n \geq N$, $N \geq 1$.

It is said that a positive operator P acting on $\mathcal{E}$ is *order continuous* if for any sequence (f_n) of Borel functions, the relation $f_n \xrightarrow{o} 0$ implies $P(f_n) \xrightarrow{o} 0$.

Definition 4.2 Let $\mu \in M(X)$ be a Borel measure on $(X, \mathcal{B})$, and let P be a positive order continuous operator on $\mathcal{F}(X, \mathcal{B})$. Define the *action* of P on the set $M(X)$ as follows: for every fixed $\mu \in M(X)$, we set

$$(\mu P)(f) := \int_X P(f)\, d\mu, \tag{4.1}$$

where f is a positive measurable function.

Applying (4.1) to characteristic functions χ_A ($A \in \mathcal{B}$), we have the formula

$$(\mu P)(A) = \int_X P(\chi_A)\, d\mu. \tag{4.2}$$

Lemma 4.3 *Suppose P is a positive order continuous operator on $\mathcal{F}(X, \mathcal{B})$. Then*

(1) μP is a sigma-finite Borel measure on $(X, \mathcal{B})$ such that

$$(\mu P)(X) < \infty \iff P(\mathbf{1}) \in L^1(\mu).$$

(2) Let X be a Polish space, and let $C_b(X)$ be the set of bounded continuous functions. If $P(C_b(X)) \subset C_b(X)$, then the action $\mu \mapsto \mu P$ is continuous with respect to the weak* topology on $M_1(X)$.

(3) The set $M_1 P := \{\mu R : \mu \in M_1(X)\}$ is a closed subset of $M_1(X)$ in the weak* topology.

Proof

(1) The assertion that μP is a measure follows directly from Definition 4.2. One can apply monotone convergence theorem, and the order continuity of P, to show that μP is countably additive.

   Since $P(\chi_A) \leq P(\mathbf{1})$, we see that finiteness of measure μP is equivalent to the property $P(\mathbf{1}) \in L^1(\mu)$.

(2) The assumption that X is a Polish space is not restrictive because any standard Borel space is Borel isomorphic to a Polish space. Thus, the set of probability

measures $M_1(X)$ can be endowed with the weak* topology. Let a sequence (μ_n) converge to a measure μ. Then, for any positive bounded function $f \in C(X)$,

$$(\mu_n P)(f) = \int_X P(f) \, d\mu_n \rightarrow \int_X P(f) \, d\mu = (\mu P)(f)$$

as $n \rightarrow \infty$.

(3) It follows from (2). □

Assumption In the sequel, we assume (sometimes implicitly) that the considered transfer operators R are order continuous. This means that we always have the well defined action of R on the set of measures $R : \mu \mapsto \mu R$. We denote $M_1(X)R = K_1$. This set of measures will play an important role in the next chapters.

In the next remark, we consider a particular case when an action of positive operators can be defined on the set of measures without additional assumptions.

Remark 4.4 Suppose X is a locally compact Hausdorff space, and P is a positive operator in the space $\mathcal{F}(X, \mathcal{B})$. Denote by $C_c(X)$ the space of continuous functions with compact support. Then, for every $x \in X$, the operator P defines a positive linear functional on $C_c(X)$ by the formula $f \mapsto P(f)(x)$. By Riesz' theorem, there exists a positive Borel measure μ^x such that

$$P(f)(x) = \int_X f \, d\mu^x. \tag{4.3}$$

Then the function $F : x \mapsto \mu^x(A)$ is measurable on X for every $A \in \mathcal{B}$ because

$$F(x) = P(\chi_A)(x).$$

Hence, for every positive operator, we can associate a measurable family of measures (μ^x) defined on $(X, \mathcal{B})$.

Now, given a measure $\nu \in M(X)$, define νP as a function on Borel sets:

$$(\nu P)(A) := \int_X P(\chi_A)(x) \, d\nu(x). \tag{4.4}$$

It follows from the definition of (μ^x) (4.3) that

$$(\nu P)(A) = \int_X \left(\int_X \chi_A \, d\mu^x \right) d\nu(x) = \int_X \mu^x(A) \, d\nu(x).$$

It is obvious that νP is a well defined complete Borel measure on $(X, \mathcal{B})$.

The concept, we considered in Remark 4.4, is known in probability theory by the name of a *random measure*. More formally, let $(X, \mathcal{B}, \mu)$ be a measure space. Then a random measure Φ defined with respect to this space is a function $x \mapsto \nu_x : X \rightarrow M(Y)$ such that $\nu_x(A)$ is $\mathcal{B}$-measurable for every $A \in \mathcal{A}$, see, e.g., [GSSY16, Sur16].

Definition 4.5 Let P be a positive operator acting in $\mathcal{F}(X, \mathcal{B})$. Given a measure $\lambda \in M(X)$, we say that P is *p-integrable* with respect to λ if $P(\mathbf{1}) \in L^p(\lambda)$. We use the terms "1-integrable" and "integrable" as synonyms.

For a fixed positive operator P, we denote

$$I_p(P) := \{\lambda \in M(X) : \int_X P(\mathbf{1})^p \, d\lambda < \infty\}.$$

It is obvious that if $\lambda_1, \lambda_2 \in I_p(R)$, then $c_1\lambda_1 + c_2\lambda_2 \in I_p(R)$.

In the next statement, we prove that the set of simple functions $\mathcal{S}(X)$ belongs to the domain of any p-integrable transfer operator R.

Lemma 4.6 *Let P be a positive operator on $\mathcal{F}(X, \mathcal{B})$. Suppose λ is a p-integrable measure. Then $P(s(x)) \in L^p(\lambda)$ for any simple function $s \in \mathcal{S}$ where $1 \leq p < \infty$.*

Proof We notice that, because P is positive ($f \leq g$ implies $P(f) \leq P(g)$), then the fact that $P(\mathbf{1})$ is in $L^p(\lambda)$ implies that $P(\chi_A)$ is in $L^p(\lambda)$. Then the statement follows from linearity of R, and we can conclude that

$$P(\mathcal{S}(X)) \subset L^p(\lambda) \iff P(\mathbf{1}) \in L^p(\lambda).$$

$\square$

Let $f \in \mathcal{F}(X, \mathcal{B})$, then by $\mathrm{supp}(f)$ we denote the Borel set $\{x \in X : f(x) \neq 0\}$.

Lemma 4.7 *Let $(X, \mathcal{B})$, R, σ be as in Definition 3.1.*

(1) For any Borel set A,

$$\mathrm{supp}(R(\chi_A)(x)) \subset \sigma(A),$$

that is $R(\chi_A)(x) = 0$ if $x \notin \sigma(A)$.
(2) If R is a strict transfer operator in $\mathcal{F}(X, \mathcal{B})$, then, for any Borel set A,

$$\mathrm{supp}(R(\chi_A)(x)) = \sigma(A),$$

$$\lambda(\sigma(A) \setminus \mathrm{supp}(R(\chi_A))) = 0.$$

(3) If λ quasi-invariant with respect to σ, $\lambda \in \mathcal{Q}_-(\sigma)$, then $\lambda(A) > 0$ implies that $\lambda(\sigma(A)) > 0$. Then statements (1) and (2) hold for λ-a.e. $x \in X$.

Proof

(1) We use the pull-out property (3.1) to prove (1). Observe that the relation $A \subset \sigma^{-1}(\sigma(A))$ holds for any endomorphism σ and any measurable set A. Then

$$\chi_A(x) = \chi_{\sigma^{-1}(\sigma(A))}(x)\chi_A(x)$$

$$= \chi_{\sigma(A)}(\sigma(x))\chi_A(x).$$

Hence,

$$R(\chi_A(x)) = R(\chi_{\sigma(A)}(\sigma(x))\chi_A(x))$$
$$= \chi_{\sigma(A)}(x)R(\chi_A(x))$$

and the result follows from

$$R(\chi_A)(x)(1 - \chi_{\sigma(A)}(x)) = 0.$$

(2) By assumption, we have that $R(\mathbf{1})(x) > 0$ for any $x \in X$. Since $\mathbf{1}(x) = \chi_A(x) + \chi_{X \setminus A}(x)$, we obtain that $R(\chi_A)(x) + R(\chi_{A^c})(x) > 0$ for all $x \in X$. It follows from (1) that

$$\text{supp}(R(\chi_A)(x)) \cap \text{supp}(R(\chi_{X \setminus A})(x)) \subset \sigma(A) \cap \sigma(X \setminus A) = \emptyset.$$

Since $R(\mathbf{1})(x) > 0$, we see that (2) holds.

(3) It follows from the relation $A \subset \sigma^{-1}(\sigma(A))$ that $\lambda(A) > 0 \implies \lambda(\sigma(A)) > 0$. Hence statement (3) is deduced from above. □

Remark 4.8 Let $\mathcal{S}(X)$ be the set of simple functions on X. Suppose R is a transfer operator acting in $L^p(\lambda)$, $1 \le p < \infty$. For $\lambda \in M(X)$, we set

$$\mathcal{S}^p(\lambda) = \mathcal{S}(X) \cap L^p(\lambda).$$

Then $\mathcal{S}^p(\lambda)$ is dense in $L^p(\lambda)$ with respect to the norm.

Corollary 4.9 *Suppose R is p-integrable transfer operator defined on the space $\mathcal{F}(X, \mathcal{B})$, $1 \le p < \infty$. Then R generates a transfer operator in $L^p(\lambda)$ with dense domain containing the set $\mathcal{S}^p(\lambda)$:*

$$R(\mathcal{S}^p(\lambda)) \subset L^p(\lambda).$$

In general, R is an unbounded linear operator in $L^p(\lambda)$.

The property of 1-integrability for a transfer operator R allows one to define a map $\lambda \mapsto \lambda R$ from the set $I_1(R) = I(R)$ to the set of *finite* measures on $(X, \mathcal{B})$.

Proposition 4.10 *Let (R, σ) be a transfer operator on $(X, \mathcal{B}, \lambda)$ such that λ is backward and forward quasi-invariant with respect to σ, i.e., $\lambda \in \mathcal{Q}_- \cap \mathcal{Q}_+$. Then, if $\lambda \in I(R)$, the relation*

$$(\lambda R)(f) := \int_X R(f)(x) \, d\lambda(x) \tag{4.5}$$

defines a finite Borel measure λR on $(X, \mathcal{B})$.

Proof By the premise of the proposition, we have $R(\mathbf{1}) \in L^1(\lambda)$. To justify relation (4.5), we use the standard approach via approximation by simple functions.

Given a λ-measurable non-negative function f, take a sequence (s_n) of simple Borel functions such that $s_n \leq s_{n+1}$ and $f(x) = \lim_n s_n(x)$ for λ-a.e. $x \in X$. Then we can define

$$\int_X R(f)(x)\, d\lambda(x) := \lim_{n \to \infty} \int_X R(s_n)(x)\, d\lambda(x), \quad f \in \mathcal{F}(X, \mathcal{B}). \qquad (4.6)$$

The limit in (4.6) exists (it may be infinite) because the sequence $(R(s_n))$ is increasing and consists of integrable functions by Lemma 4.7. In fact, we can see that the definition in (4.6) does not depend on the choice of a sequence (s_n) since

$$\lim_{n \to \infty} \int_X R(s_n)(x)\, d\lambda(x) = \sup \left(\int_X R(s)(x)\, d\lambda(x) \,:\, s(x) \leq f(x), \forall x \in X \right).$$

We need to show relation (4.5) defines a Borel measure on $(X, \mathcal{B})$. Set, for any $A \in \mathcal{B}$,

$$(\lambda R)(A) := \lambda(R(\chi_A)).$$

To see that λR is sigma-additive, it suffices to prove that if $A_i \supset A_{i+1}$ and $\bigcap_i A_i = \emptyset$, then

$$\lim_{i \to \infty} (\lambda R)(A_i) = 0. \qquad (4.7)$$

It follows from Lemma 4.7 that $(\lambda R)(A_i) \leq \lambda(\sigma(A_i))$. Since σ is forward quasi-invariant, we see that $\lambda(\sigma A_i) \to 0$ as $i \to \infty$. Hence, relation (4.7) holds, and λR is a sigma-additive measure.

Moreover, we see from the equality

$$(\lambda R)(X) = \int_X R(\mathbf{1})\, d\lambda,$$

that λR is a finite measure. $\square$

In general, the measure λR is not absolutely continuous with respect to λ, see e.g., Example 1.2 and Table 1.1. One of our aims is to find conditions under which $\lambda R \ll \lambda$. We recall the following example (more details are in [DJ06, Section 4]).

Example 4.11 (Case of Wavelets [DJ06], See Also Example 1.2) Let $\mathbb{T} = \{z \in \mathbb{C} : |z| = 1\}$ be the unit circle, and let $\sigma_2(z) = z^2$, $\sigma_3(z) = z^3$ be two surjective endomorphisms of $\mathbb{T}$. Suppose the transfer operator R_i on $C(\mathbb{T})$ is defined as follows:

$$R_i(f)(z) = \sum_{w:\sigma_i(w)=z} |m(w)|^2 f(w), \quad i = 2, 3,$$

where $m(w) = \dfrac{1 + w^2}{\sqrt{2}}$. It was proved in [DJ06] that:

(a) the measure δ_1 (the Dirac point mass measure at $z = 1$) is R_i-invariant, $\delta_1 R_i = \delta_1$;
(b) for the transfer operator (R_3, σ_3), the Riesz measure

$$d\nu(t) = \lim_{n \to \infty} \frac{1}{2\pi} \prod_{k=1}^{n} (1 + \cos(2 \cdot 3^k t))$$

is singular with respect to the Lebesgue measure, and satisfies the relation $\nu R_3 = \nu$.

The reader can find more results on wavelets in the cited above [BJ02, DJ06] and [AJL16].

Remark 4.12 We are interested in the following problem: Given a transfer operator (R, σ) acting in $\mathcal{F}(X, \mathcal{B})$, find a Borel measure λ such that $\lambda R \ll \lambda$. We showed in Examples 1.2, and 4.11 that a transfer operator may satisfy, or not satisfy, this condition of absolute continuity. In order to formulate the problem correctly, an action of R on measures must be well defined. We know that this is always the case when R is *order continuous* (see Lemma 4.3), or when X is a locally compact Hausdorff space (see Remark 4.4). On the other hand, if we restrict our choice of measures to the subsets of σ-quasi-invariant measures and integrable functions $R(1)$, we still have a vast set of measures which includes interesting applications. We note that if R is a normalized transfer operator, then the latter condition automatically holds for all finite measures.

Based on these observations, we will assume that, for a transfer operator R, the map $\lambda \mapsto \lambda R$ is defined on $M(X)$ (or on a subset of $M(X)$ in case of need).

Definition 4.13 For a fixed order-continuous transfer operator (R, σ), we define the set

$$\mathcal{L}(R) := \{\lambda \in M(X) : \lambda R \ll \lambda\}. \tag{4.8}$$

In case when R is integrable, we use the same notation $\mathcal{L}(R)$ for the set of Borel measures λ such that $\lambda R \ll \lambda$, $R(\mathbf{1}) \in L^1(\lambda)$, and λ is quasi-invariant with respect to σ.

We are interested in the following questions.

Question

(1) Under what conditions on (R, σ) is the set $\mathcal{L}(R)$ non-empty?
(2) What properties does the set $\mathcal{L}(R)$ have? In particular, can we iterate the map $\lambda \mapsto \lambda R$ infinitely many times?

We first give a partial answer to Question (1) in the following theorem.

Theorem 4.14

(1) Let (R, σ) be a transfer operator defined on a standard measure space $(X, \mathcal{B}, \lambda)$ such that $R(\mathbf{1}) \in L^1(\lambda)$. Then

$$\mathcal{L}(R) \supset I(R) \cap \mathcal{Q}_+.$$

In other words, if σ is backward and forward quasi-invariant with respect to λ, and $\lambda \in I(R)$, then $\lambda R \ll \lambda$.

(2) If, additionally to the conditions in (1), we assume that (R, σ) is a strict transfer operator, then λR is equivalent to λ and

$$\{\lambda : \lambda R \sim \lambda\} = I(R) \cap \mathcal{Q}_- \cap \mathcal{Q}_+.$$

Proof

(1) Let A be a Borel subset of X. By Lemma 4.7, the function $R(\chi_A)$ is non-negative and integrable with respect to λ. As was shown in the proof of Lemma 4.7, the relation

$$\chi_A(x) = \chi_{\sigma(A)}(\sigma(x))\chi_A(x)$$

holds. Based on this fact and the pull-out property for R, we calculate

$$
\begin{aligned}
(\lambda R)(A) &= \int_X R(\chi_A)\, d\lambda(x) \\
&= \int_X R[\chi_{\sigma(A)}(\sigma(x))\chi_A(x)]\, d\lambda(x) \qquad (4.9) \\
&= \int_X \chi_{\sigma(A)}(x) R(\chi_A)(x)\, d\lambda(x) \\
&= \int_{\sigma(A)} R(\chi_A)(x)\, d\lambda(x).
\end{aligned}
$$

Now if $\lambda(A) = 0$, then, by the assumption, $\lambda(\sigma(A)) = 0$. Therefore $\lambda R(A) = 0$, and this proves that $\lambda R \ll \lambda$.

(2) Having statement (1) proved, we need to show that $\lambda(A) > 0$ implies that $(\lambda R)(A) > 0$. As $A \subset \sigma^{-1}(\sigma(A))$, and σ is backward non-singular, we obtain that $\lambda(A) > 0 \implies \lambda(\sigma(A)) > 0$. In this case, we see from (4.9) that

$$\int_{\sigma(A)} R(\chi_A)(x)\, d\lambda(x) > 0$$

because of Lemma 4.7. Thus, it follows from (4.9) that $(\lambda R)(A) > 0$. $\qquad\square$

Remark 4.15 Note that the condition $\lambda \circ R \ll \lambda$ is based on the forward quasi-invariance of σ. On the other hand, If we additionally assume that R is a strict transfer operator, then our proof of $\lambda \ll \lambda \circ R$ is based on the backward quasi-invariance of σ. In fact, the condition that R is a strict transfer operator can be slightly weakened as shown in the next statement.

Corollary 4.16 *Let (R, σ) be as in Theorem 4.14 and $\lambda \in I(R)$. Suppose that for every measurable set A with $\lambda(A) > 0$, the function $R(\chi_A)$ is nonzero as a function in $L^1(\lambda)$. Then the measures λ and λR are equivalent.*

Proof It suffices to show that $(\lambda R)(A) = 0$ implies that $\lambda(A) = 0$ because the converse result was proved in Theorem 4.14. We use (4.9) and the fact that the support of the function $R(\chi_A)$ belongs to $\sigma(A)$. Hence, if $\lambda(A) > 0$, then, by quasi-invariance of σ, we conclude that $\lambda(\sigma(A)) > 0$. Therefore, by the assumption,

$$\int_{\sigma(A)} R(\chi_A) \, d\lambda > 0.$$

and finally we obtain that $(\lambda R)(A) > 0$. □

The following result states that the measures λR and λ are equivalent when they are restricted on the sigma-subalgebra $\sigma^{-1}(\mathcal{B})$ of $\mathcal{B}$. This result agrees with the fact proved in the setting of Borel dynamics. Namely, we showed in Lemma 3.11 that R is a one-to-one map from $\mathcal{F}(X, \sigma^{-1}(\mathcal{B}))$ onto $\mathcal{F}(X, \mathcal{B})$.

Proposition 4.17 *Suppose that (R, σ) is a transfer operator on a standard Borel space $(X, \mathcal{B})$, and let $\lambda \in I(R)$. If σ is a non-singular endomorphism on $(X, \mathcal{B})$ with respect to λ, then*

$$\lambda R|_{\sigma^{-1}(\mathcal{B})} \sim \lambda|_{\sigma^{-1}(\mathcal{B})}.$$

Proof For any set $A \in \mathcal{B}$, set $B = \sigma^{-1}(A)$. By non-singularity of σ, we have

$$(\lambda(B) = 0) \iff (\lambda(A) = 0).$$

On the other hand, we can apply the same method as in Theorem 4.14 and obtain that

$$(\lambda R)(B) = \int_X R(\chi_B)(x) \, d\lambda(x)$$

$$= \int_X R(\chi_{\sigma^{-1}(A)})(x) \, d\lambda(x)$$

$$= \int_X R((\chi_A \circ \sigma)(x)) \, d\lambda(x)$$

$$= \int_X \chi_A(x) R(\mathbf{1})(x) \, d\lambda(x)$$

$$= \int_A R(\mathbf{1})(x) \, d\lambda(x).$$

Since $R(\mathbf{1}) \in L^1(X, \mathcal{B}, \lambda)$, we conclude that $(\lambda R)(B) = 0$ if and only if $\lambda(A) = 0$ if and only if $\lambda(B) = 0$. $\square$

4.2   Ergodic Decomposition of Transfer Operators

Let $(X, \mathcal{B})$ be a standard Borel space. Fix a probability Borel measure $\lambda \in M(X)$ and consider the dynamical system $(X, \mathcal{B}, \lambda, \sigma)$ where σ is an onto endomorphism of X such that the measure λ is σ-quasi-invariant. The fact that every dynamical system can be decomposed into ergodic components is extremely important in measurable dynamics because it allows to reduce many problems to the case of ergodic transformations. We aim to establish a similar fact for arbitrary transfer operator. It will be shown that if ξ is a measurable partition invariant with respect to σ, then every transfer operator (R, σ) defines a *measurable field* of transfer operators. Then this result is applied to the partition into ergodic components for σ.

In what follows, we describe the structure of measurable fields of endomorphisms and transfer operators with respect to a measurable partition.

Suppose that ξ is a measurable partition of $(X, \mathcal{B}, \lambda)$ which is σ-invariant, i.e., $\sigma^{-1}(C) = C$ for any element C of the partition ξ. Then the quotient space $(X/\xi, \mathcal{B}/\xi, \lambda_\xi)$ is a standard probability measure space. Moreover, there exists a system of conditional measures $C \mapsto \lambda_C, C \in X/\xi$ such that $(C, \mathcal{B} \cap C, \lambda_C)$ is a standard probability measure space, the function $C \mapsto \lambda_C(A \cap C)$ is $\mathcal{B}/\xi$-measurable for any $A \in \mathcal{B}$, and

$$\lambda(f) = \int_{X/\xi} \lambda_C(f) \, d\lambda_\xi(C),$$

see Sect. 2.3 and Definition 2.7 for more details.

The fact that ξ is σ-invariant means that the endomorphism σ generates a measurable field $C \mapsto \sigma_C$ of surjective endomorphisms σ_C of $(C, \mathcal{B}_C, \lambda_C)$ such that λ_C is σ_C-quasi-invariant. Conversely, one can begin with a measurable field of endomorphisms $C \mapsto \sigma_C$ to define an endomorphism σ of $(X, \mathcal{B}, \lambda)$.

Since the measure spaces $(X/\xi, \mathcal{B}/\xi, \lambda_\xi)$ and $(C, \mathcal{B} \cap C, \lambda_C), C \in X/\xi$ are standard, we can simplify our notation by applying a Borel isomorphism Φ between the following measure spaces:

$$\Phi : (X, \mathcal{B}, \lambda) \rightarrow (Y \times Z, \mathcal{D} \times \mathcal{A}, \mu \times \nu),$$

where $(Y, \mathcal{D}, \mu)$ is isomorphic to $(X/\xi, \mathcal{B}/\xi, \lambda_\xi)$ and $(Z, \mathcal{A}, \nu)$ is isomorphic to $(C, \mathcal{B} \cap C, \lambda_C)$ for every $C \in X/\xi$. Using this isomorphism Φ, we transfer the field of endomorphisms $C \mapsto \sigma_C$ to the space $Y \times Z$ and denote it as $y \mapsto \sigma_y$.

Suppose now that a *measurable field of transfer operators* $y \mapsto (R_y, \sigma_y)$ is given. This means that $R_y : \mathcal{F}(Z, \mathcal{A}) \to \mathcal{F}(Z, \mathcal{A})$ satisfies the conditions

$$R_y((f \circ \sigma_y)g) = f R_y(g), \tag{4.10}$$

and the function

$$(y, z) \mapsto R_y(f_y)(z) \tag{4.11}$$

is Borel where $f_y(z) = f(y, z) \in \mathcal{F}(Z, \mathcal{A})$, and f is Borel on $(Y \times Z, \mathcal{D} \times \mathcal{A})$.

Lemma 4.18 *Suppose that ξ is a measurable partition invariant with respect to a surjective endomorphism σ. Let $y \mapsto (R_y, \sigma_y)$ be a measurable field of transfer operators defined as above. Then the linear operator R acting in $\mathcal{F}(Y \times Z, \mathcal{D} \times \mathcal{A})$ by the formula*

$$R(f)(y, z) = R_y(f_y)(z) \tag{4.12}$$

is a transfer operator with respect to σ.

Conversely, every transfer operator (R, σ) on $\mathcal{F}(Y \times Z, \mathcal{D} \times \mathcal{A})$ generates a measurable field of transfer operators $y \mapsto (R_y, \sigma_y)$.

Proof The operator R is well defined due to relation (4.11). We use (4.10) and (4.12) to show that (R, σ) satisfies the pull-out property: for $f, g \in \mathcal{F}(Y \times Z, \mathcal{D} \times \mathcal{A})$,

$$R((f \circ \sigma)g)(y, z) = R_y((f_y \circ \sigma_y)g_y)(z)$$

$$= f_y(z) R_y(g_y)(z) \tag{4.13}$$

$$= f(y, z) R(g)(y, z).$$

The converse statement is obvious. □

Suppose that (R, σ) is a transfer operator in $\mathcal{F}(X, \mathcal{B})$. We define now an analogue of ergodic decomposition for a transfer operator (R, σ).

We recall the fact that follows from Lemma 3.5: if A is a σ-invariant set, then the restriction of R to the space of Borel functions on A gives an R_A. It is well known that if σ is non-ergodic, then one can use the standard procedure of ergodic decomposition for σ, see e.g. [CFS82]. We show here that, in this case, any transfer operator R related to σ also admits an ergodic decomposition.

Let $(X, \mathcal{B}, \lambda, \sigma)$ be a non-singular non-ergodic dynamical system. Without loss of generality, we can assume that the measure λ is probability. Consider the partition ζ of X into orbits of σ. We recall that, by definition, x and y are in the same *orbit* of σ if there are positive integers n, m such $\sigma^n(x) = \sigma^m(y)$. Let ξ be the measurable hull of ζ, i.e., ξ is a maximal measurable partition with property $\xi \prec \zeta$, that is every element C of ξ is a ζ-set. We observe that ξ is a trivial partition in the case when σ is ergodic. Obviously, ξ is σ-invariant. Denote by $(X/\xi, \mathcal{B}/\xi, \lambda_\xi)$ the quotient measure space. By Theorem 2.8, there exists a unique system of conditional measures $(\lambda_C)_{C \in X/\xi}$ such that

$$\lambda(A) = \int_{X/\xi} \lambda_C(B) \, d\lambda_\xi(C),$$

and, for any measurable function f on X,

$$\int_X f \, d\lambda = \int_{X/\xi} \left(\int_C \chi_C f \, d\lambda_C \right) d\lambda_\xi(C). \tag{4.14}$$

Here C is an arbitrary element of the partition ξ and can be considered as a point in X/ξ. We refer to Sect. 2.3 for more information about conditional measures.

Now we apply the result proved in Lemma 4.18. Let (R, σ) be a transfer operator in $\mathcal{F}(X, \mathcal{B})$ and $\lambda \in I(R)$ be a fixed measure. Suppose that $\lambda \in \mathcal{L}(R)$. Then there exists an integrable measurable function W such that

$$\int_X R(f) \, d\lambda = \int_X f W \, d\lambda, \tag{4.15}$$

i.e., $W(x) = \dfrac{d(\lambda R)}{d\lambda}(x)$ is the *Radon-Nikodym derivative*.

Theorem 4.19 *Let $(X, \mathcal{B}, \lambda, \sigma)$ be as above and $\lambda(X) = 1$. Let ξ be the partition into ergodic components of the endomorphism σ. Suppose (R, σ) is a transfer operator defined on the space $L^1(\lambda)$ such that $\lambda R \ll \lambda$. Let (λ_C) be the system of conditional measures defined by the measurable partition ξ on the measure space $(X, \mathcal{B}, \lambda)$. Then the operator R is decomposed into a measurable field of transfer operators (R_C, σ_C) such that $\lambda R_C \ll \lambda_C$, and*

$$W_C := \frac{d(\lambda_C R_C)}{d\lambda_C} = W\chi_C,$$

where $W d\lambda = d(\lambda R)$.

Proof Let $(C, \mathcal{B} \cap C, \lambda_C)$ be the standard measure space obtained by restriction of Borel sets to $C \in X/\xi$. Then C is σ-invariant, and, by Lemma 3.5, we can define a transfer operator (R_C, σ) by the formula $R_C(f) = R(f)|_C$. It follows from (4.14) that R_C is a transfer operator in the space $L^1(\lambda_C)$. Then we can compute

$$\int_X R(f) \, d\lambda = \int_X f W \, d\lambda$$

$$= \int_{X/\xi} \left(\int_C f W \chi_C \, d\lambda_C \right) d\lambda_\xi$$

On the other hand, we have

$$\int_X R(f) \, d\lambda = \int_{X/\xi} \left(\int_C R(f) \chi_C \, d\lambda_C \right) d\lambda_\xi$$

$$= \int_{X/\xi} \left(\int_C R[f(\chi_C \circ \sigma)] \, d\lambda_C \right) d\lambda_\xi$$

$$= \int_{X/\xi} \left(\int_C f \chi_C \, d(\lambda_C R_C) \right) d\lambda_\xi$$

$$= \int_{X/\xi} \left(\int_C f W_C \, d\lambda_C \right) d\lambda_\xi.$$

By uniqueness of the system of conditional measures, we have the result.   □

4.3 Positive Operators and Polymorphisms

In this section, we discuss the following *problem*. Let (X, B) be a standard
Borel space, and let μ_1, μ_2 be two probability measures on (X, B). How can we
characterize positive operators P such that $\mu_1 P = \mu_2$? What can be said about
transfer operators (R, σ) satisfying the condition $\mu_1 R = \mu_2$ where σ is an onto
endomorphism of (X, B)?

We denote by $\mathcal{P}$ the set of positive linear operators acting in the space of
Borel functions $\mathcal{F}(X, B)$. We assume implicitly that operators from $\mathcal{P}$ are order
continuous so that the action $\mu \mapsto \mu P$ is defined on $M(X)$.

For fixed measures μ_1 and μ_2 on (X, B), we denote

$$\mathcal{P}(\mu_1, \mu_2) := \{P \in \mathcal{P} : \mu_1 P = \mu_2, \ P(\mathbf{1}) = \mathbf{1}\}.$$

Similarly, if $\mathcal{R}(\sigma)$ is the set of transfer operators corresponding to an endomorphism
σ, then we denote

$$\mathcal{R}(\mu_1, \mu_2) := \{R \in \mathcal{R}(\sigma) : \mu_1 R = \mu_2, \ R(\mathbf{1}) = \mathbf{1}\}.$$

Given a standard Borel space (X, B), consider the product space $(Y, A) = (X \times
X, B \times B)$, and let π_1 and π_2 be the projections, $\pi_i(x_1, x_2) = x_i, i = 1, 2$. For
convenience of notation, we will also write $Y = X_1 \times X_2$ where $X_1 = X = X_2$. It
will be clear from our next discussions that all results remain true in the case when
we have two distinct spaces (X_1, B_1, μ_1) and (X_2, B_2, μ_2).

Suppose that ν is a Borel probability measure on $X \times X$. Then ν defines the
marginal measures μ_1 and μ_2 on X_1 and X_2, respectively:

$$\mu_i(A) = \nu(\pi_i^{-1}(A)), \quad A \in B.$$

Denote by

$$\mathfrak{M}(\mu_1, \mu_2) := \{\nu \in M_1(Y) : \nu \circ \pi_1^{-1} = \mu_1, \ \nu \circ \pi_2^{-1} = \mu_2\}.$$

We remark that if two measures μ_1, μ_2 are given on X, then the product measure $\nu = \mu_1 \times \mu_2$ gives an example of a measure from $\mathfrak{M}(\mu_1, \mu_2)$.

Lemma 4.20 *Suppose ν is a probability measure on $Y = X_1 \times X_2$ from the set $\mathfrak{M}(\mu_1, \mu_2)$. Then ν is uniquely determined by the system of conditional measures $(\nu_x : x \in X_1)$ generated by the measurable partition $\xi_1 := \{\pi_1^{-1}(x) : x \in X_1\}$,*

$$\nu = \int_{X_1} \nu_x \, d\mu_1(x)$$

This result follows immediately from the uniqueness of the system of conditional measures (see Sect. 2.3). It can be interpreted as follows: if ν and ν' are two measures from $\mathfrak{M}(\mu_1, \mu_2)$, and $\nu_x = \nu'_x$ for μ_1-a.e. $x \in X_1$, then $\nu = \nu'$.

For every measure $\nu \in \mathfrak{M}(\mu_1, \mu_2)$, we define a positive operator $P_\nu : L^1(\mu_2) \to L^1(\mu_1)$ by setting

$$P_\nu(f)(x) = \mathbb{E}_\nu(f \circ \pi_2 \mid \pi_1^{-1}(x)). \tag{4.16}$$

Equivalently, formula (4.16) can be written as follows

$$P_\nu(f)(x) = \int_{X_2} (f \circ \pi_2) \, d\nu_x, \quad x \in X_1, \tag{4.17}$$

where ν_x is the system of conditional measures defined in Lemma 4.20.

We observe that if $\nu = \mu_1 \times \mu_2$, then P_ν is a rank 1 operator such that

$$P_\nu(f) = \int_{X_2} (f \circ \pi_2) \, d\mu_2,$$

so that $P_\nu(f)(x), x \in X_1$, is a constant function.

For the measure space $(Y, \mathcal{A}, \nu) = (X_1 \times X_2, \mathcal{B} \times \mathcal{B}, \nu)$, the projections π_1 and π_2 define the *isometries* V_1 and V_2, respectively, where

$$V_1(f) = f \times 1 : L^2(X_1, \mu_1) \to L^2(Y, \nu), \tag{4.18}$$

$$V_2(f) = 1 \times f : L^2(X_2, \mu_2) \to L^2(Y, \nu). \tag{4.19}$$

With some abuse of notation, we denote by $f_1 \times f_2$ the function $f(x_1, x_2) = f_1(x_1) f_2(x_2)$. Equivalently, $(f \times 1)(x_1, x_2) = f \circ \pi_1(x_1, x_2)$, and $(1 \times f)(x_1, x_2) = f \circ \pi_2(x_1, x_2)$.

Lemma 4.21 *The operator P_ν, considered as an operator acting from $L^2(\mu_2)$ into $L^2(\mu_1)$, satisfies the relation*

$$P_\nu = V_1^* V_2.$$

Proof We begin with finding the explicit formula for the adjoint operator V_1^* : $L^2(\nu) \to L^2(\mu_1)$. For any functions $f \in L^2(\nu)$ and $g \in L^2(\mu_1)$, we have

$$\langle f, V_1(g) \rangle_{L^2(\nu)} = \int_{X_1 \times X_2} f(x_1, x_2)(g \circ \pi_1)(x_1, x_2)\, d\nu(x_1, x_2)$$

$$= \int_{X_1} (g \circ \pi_1)(x_1, x_2) \left(\int_{X_2} f(x_1, x_2)\, d\nu_{x_1} \right) d\mu(x_1)$$

$$= \langle V_1^*(f), g \rangle_{L^2(\mu_1)},$$

where

$$V_1^*(f)(x_1) = \int_{X_2} f(x_1, x_2)\, d\nu_{x_1}. \tag{4.20}$$

The remaining part of the proof follows now from (4.20):

$$V_1^* V_2(f) = V_1^*(f \circ \pi_2) = \int_{X_2} (f \circ \pi_2)(x_1, x_2)\, d\nu_{x_1}(x_2) = P_\nu(f),$$

and we are done.   □

Our main result of this section, Theorem 4.22, contains several statements that clarify the relationship between the set of measures $\mathfrak{M}(\mu_1, \mu_2)$ and the set of positive operators $P \in \mathcal{P}(\mu_1, \mu_2)$. We use here the notation introduced above. We also consider positive operators acting in the corresponding L^2-spaces.

Theorem 4.22

(1) Let $\nu \in \mathfrak{M}(\mu_1, \mu_2)$. Then formula (4.16) defines an affine map

$$\Psi(\nu) = P_\nu : \mathfrak{M}(\mu_1, \mu_2) \to \mathcal{P}(\mu_1, \mu_2).$$

(2) Let $P \in \mathcal{P}(\mu_1, \mu_2)$ be a positive operator acting in $\mathcal{F}(X, \mathcal{B})$. Define a measure ν_P on $(X \times X, \mathcal{B} \times \mathcal{B})$ by setting

$$\nu_P(f_1 \times f_2) := \int_X f_1 P(f_2)\, d\mu_1. \tag{4.21}$$

Then, $\Phi(P) = \nu_P$ defines an affine map

$$\Phi : \mathcal{P}(\mu_1, \mu_2) \to \mathfrak{M}(\mu_1, \mu_2).$$

(3) *The maps* $\Psi : v \mapsto P_v$ *and* $\Phi : P \mapsto v_P$ *are affine bijections between the sets* $\mathfrak{M}(\mu_1, \mu_2)$ *and* $\mathcal{P}(\mu_1, \mu_2)$ *such that* $\Psi \circ \Phi(P) = P$, *and* $\Phi \circ \Psi(v) = v$.

Proof

(1) We first check that the operator P_v belongs to the set $\mathcal{P}(\mu_1, \mu_2)$. Since v is a probability measure, P_v is a normalized positive operator, we obtain

$$
\int_{X_1} P_v(f)(x_1) \, d\mu_1(x_1) = \int_{X_1} \left(\int_{X_2} (f \circ \pi_2)(x_1, x_2) \, dv_{x_1}(x_2) \right) d\mu_1(x_1)
$$

$$
= \int_{X_1 \times X_2} (f \circ \pi_2)(x_1, x_2) \, dv(x_1, x_2)
$$

$$
= \int_{X_2} \left(\int_{X_1} (f \circ \pi_2)(x_1, x_2) \, dv_{x_2}(x_1) \right) d\mu_2(x_2)
$$

$$
= \int_{X_2} f(x_2) \, d\mu_2(x_2).
$$

We used here the Fubini theorem, and two properties: (1) the measure v_x is probability for μ_1-a.e $x \in X_1$, and (2) $X_1 = X_2 = X$. Thus, we conclude that $\mu_1 P_v = \mu_2$.

Moreover, as we will see from (2), a positive normalized operator P in $\mathcal{F}(X, \mathcal{B})$ belongs to $\mathcal{P}(\mu_1, \mu_2)$ if and only if there exists a measure $v \in \mathfrak{M}(\mu_1, \mu_2)$ such that $P = P_v$ where P_v is defined by (4.16).

The fact that $\Psi(\alpha v_1 + (1 - \alpha)v_2) = \alpha\Psi(v_1) + (1 - \alpha)\Psi(v_2), \alpha \in (0, 1)$, is obvious.

(2) We show that $v_P \in \mathfrak{M}(\mu_1, \mu_2)$. Apply the definition of v_P to characteristic functions χ_A and χ_B where $A \subset X_1$ and $B \subset X_2$:

$$
v_P(\chi_{A \times B}) = \int_X \chi_A P(\chi_B) \, d\mu_1.
$$

Then we see that $v \circ \pi_1^{-1}(A) = \mu_1$, and $v \circ \pi_2^{-1}(B) = \mu_1 P$. Because $P \in \mathcal{P}(\mu_1, \mu_2)$ we have $\mu_1 P = \mu_2$. Hence, μ_1 and μ_2 are the marginal measures for v_P.

Next, we show that Φ is a one-to-one map. Suppose there are positive operators $P, Q \in \mathcal{P}(\mu_1, \mu_2)$ such that $v_P = v_Q$. Then, for any functions f_1 and f_2, we have

$$
\int_X f_1 P(f_2) \, d\mu_1 = \int_X f_1 Q(f_2) \, d\mu_1.
$$

It follows, by standard arguments, that $P = Q$.

(3) It remains to check that Ψ and Φ are inverses of each other. It can be done by direct computations:

$$
\begin{aligned}
(\Phi \circ \Psi(v))(f_1 \times f_2) &= \int_{X_1} f_1 P_v(f_2) \, d\mu_1 \\
&= \int_{X_1} f_1(x_1) \left(\int_{X_2} (f_2 \circ \pi_2)(x_1, x_2) \, dv_{x_1}(x_2) \right) d\mu_1(x_1) \\
&= \int_{X_1} \int_{X_2} f_1(x_1) f_2(x_2) \, dv(x_1, x_2) \\
&= v(f_1 \times f_2).
\end{aligned}
$$

Similarly, we can show that $\Psi \circ \Phi(P) = P$ for any $P \in \mathcal{P}(\mu_1, \mu_2)$. This is equivalent to the equality $P_{v_P} = P$ where v_P is defined by (4.21). The latter relation can be proved by using the definitions of Φ and Ψ.   □

Remark 4.23 Since the maps Φ and Ψ are affine, we obtain from Theorem 4.22 that they establish one-to-one correspondence between extreme points of the sets $\mathfrak{M}(\mu_1, \mu_2)$ and $\mathcal{P}(\mu_1, \mu_2)$. We observe that the measure $\mu_1 \times \mu_2$ is an extreme point in $\mathfrak{M}(\mu_1, \mu_2)$ as well as the corresponding rank one operator $P_{\mu_1 \times \mu_2}$ is an extreme point in $\mathcal{P}(\mu_1, \mu_2)$.

Corollary 4.24 *Suppose the L^2-space of $(X_1 \times X_2, \mathcal{B} \times \mathcal{B}, v)$, $v \in \mathfrak{M}(\mu_1, \mu_2)$, is isometrically embedded into $L^2(\Omega, \rho)$ where (Ω, ρ) is a standard measure space, $U : L^2(X_1 \times X_2, v) \to L^2(\Omega, \rho)$. Let $V_i : L^2(X_i, \mu_i) \to L^2(X_1 \times X_2, v)$, $i = 1, 2$, be the isometries defined by (4.18) and (4.19). Set $\widetilde{V}_i = U V_i$. Then the positive operator $\widetilde{P}$ defined by $\widetilde{V}_i$ as in Lemma 4.21 coincides with P_v.*

The proof follows immediately from the relation

$$
\widetilde{P} = \widetilde{V}_1^* \widetilde{V}_2 = V_1^* U^* U V_2 = P_v.
$$

Suppose now an onto endomorphism σ is defined on a standard Borel space $(X, \mathcal{B})$. Let μ be a probability measure on $(X, \mathcal{B})$. We know that the partition $\xi = \{\sigma^{-1}(x) : x \in X\}$ of X is measurable, hence there exists a system of conditional measures (μ_{C_x}) defined by ξ, where C_x is the element of ξ that contains x, see Sect. 2.3. In Example 1.6, we used measures (μ_{C_x}) to define a transfer operator

$$
R(f)(x) = \int_{C_x} f(y) \, d\mu_{C_x}(y). \tag{4.22}
$$

We consider here another class of measures on the product space $(X \times X, \mathcal{B} \times \mathcal{B})$ associated to $(X, \mathcal{B}, \mu, \sigma)$. For μ, σ, and (μ_{C_x}) as above, take the partition ξ_1 of $X \times X$ into the fibers $\{\pi_1^{-1}(x) : x \in X\}$ and assign the measure μ_{C_x} to the set $\{x\} \times \pi_1^{-1}(x)$ endowed the induced Borel structure. We see that, in fact, the measure μ_{C_x} is supported by the set $\{x\} \times C_x$.

Let now v be the measure on $(X \times X, \mathcal{B} \times \mathcal{B})$ such that, for a Borel function $f(x_1, x_2)$,

$$v(f) = \int_{X_1} \left(\int_{\pi_1^{-1}(x_1)} f(x_1, x_2) \, d\mu_{C_{x_1}}(x_2) \right) d\mu(x_1). \tag{4.23}$$

Using the partition ξ_1, we can also disintegrate v over X_1 and get the family of conditional measures $v_x, x \in X_1$. By definition of v, we have $v_x = \mu_{C_x}$.

Lemma 4.25 *Let R and v be defined by (4.22) and (4.23), respectively. Then $R = R_v$ where R_v is defined in (4.16)*

(We can apply the results proved above for the operators P and P_v since R is also a positive operator.)

Proof We have from the definition of v

$$R_v(f)(x) = \mathbb{E}_v(f(\pi_2(x, y)) \mid \pi_1^{-1}(x))$$

$$= \int_{\pi_1^{-1}(x)} f(y) \, dv_x(y)$$

$$= \int_{C_x} f(y) \, d\mu_x(y)$$

$$= R(f)(x), \qquad (x, y) \in X \times X.$$

$\square$

This means, in particular, that the transfer operator R_v possesses the pull-out property.

Proposition 4.26 *Let the measure v on $X \times X$ be defined by (4.23). Then the marginal measures for v are $\mu_1 = v \circ \pi_1^{-1} = \mu$ and $\mu_2 = v \circ \pi_2^{-1}$ where*

$$\mu_2(B) = \int_X \mu_{C_x}(B) \, d\mu(x), \quad B \in \mathcal{B}.$$

Moreover, $\mu_1 R_v = \mu_2$ and $R_v \in \mathcal{R}(\mu_1, \mu_2)$.

Proof The fact that the marginal measure μ_1 coincides with μ is obvious. To find $\mu_2(B) = v(X_1 \times B)$, we take

$$\mu_2(B) = \int_X \left(\int_{\pi_1^{-1}(x)} \chi_B(y) \, d\mu_{C_x}(y) \right) d\mu(x)$$

$$= \int_X \mu_{C_x}(B) \, d\mu(x).$$

The second statement is a reformulation of the first result. Indeed, if we use (4.22) and the relation $R_\nu = R$, then we can conclude that

$$\mu_2(f) = \int_{X_1} \mu_{C_x}(f) \, d\mu(x)$$

$$= \int_{X_1} R(f) \, d\mu_1(x).$$

This completes the proof.   □

We give one more example of a measure ν on the product space $(X \times X, \mathcal{B} \times \mathcal{B})$. Let μ be a measure on $(X, \mathcal{B})$, and let σ be an onto endomorphism of X. Define the probability measure $\nu = \nu(\sigma)$ on $\mathcal{B} \times \mathcal{B}$ as follows:

$$\nu(\sigma)(A \times B) := \mu(A \cap \sigma^{-1}(B)), \quad A, B \in \mathcal{B}. \tag{4.24}$$

Lemma 4.27 *In the above notation, the following properties hold:*

(1) For $\nu = \nu(\sigma)$,

$$\mu_1 = \nu \circ \pi_1^{-1} = \mu, \quad \mu_2 = \nu \circ \pi_2^{-1} = \mu \circ \sigma^{-1}.$$

(2) The composition operator $S_\sigma : f \mapsto f \circ \sigma$ belongs to $\mathcal{P}(\mu_1, \mu_2)$ where $\mu_2 = \mu_1 S_\sigma, \mu_1 = \mu$, i.e.,

$$\mu S_\sigma = \mu \circ \sigma^{-1}.$$

(3) Let $P_{\nu(\sigma)}$ be the positive operator defined by the measure $\nu(\sigma)$ according to (4.16). Then $P_{\nu(\sigma)} = S_\sigma$. Equivalently, the operator S_σ is the only solution to the equation

$$\nu(\sigma)(f_1 \times f_2) = \int_{X_1} f_1 S(f_2) \, d\mu_1. \tag{4.25}$$

Proof The first two assertions of this lemma are rather obvious: (1) follows immediately from the definition of $\nu(\sigma)$, and (2) is verified straightforward.

To see that (3) holds, we note that, by (1), $\nu(\sigma) \in \mathfrak{M}(\mu, \mu \circ \sigma^{-1})$, and therefore, we can use Theorem 4.22. Since the maps Φ and Ψ are one-to-one, we conclude that there exists only one operator satisfying (4.25).   □

Remark 4.28

(1) There is a clear connection between positive operators from the set $\mathcal{P}(\mu_1, \mu_2)$ and the notion of *polymorphisms* which was introduced and studied in a series of papers by A. Vershik, see e.g., [Ver00, Ver05].

By definition, a polymorphism Π of the standard measure space $(X, \mathcal{B}, \mu)$ to itself is a diagram consisting of an ordered triple of standard measure spaces:

$$(X, \mathcal{B}, \mu_1) \xleftarrow{\pi_1} (X \times X, \mathcal{B} \times \mathcal{B}, \nu) \xrightarrow{\pi_2} (X, \mathcal{B}, \mu_2),$$

where π_1 and π_2 are the projections to the first and second component of the product space $(X \times X, \mathcal{B} \times \mathcal{B}, \nu)$ such that $\nu \circ \pi_i^{-1} = \mu_i$.

This definition can be naturally extended to the case of two different measure spaces $(X_i, \mathcal{B}_i, \mu_i), i = 1, 2$. Then we have a polymorphism defined between these measure spaces.

(2) Our approach to the study of measures on product spaces is similar to the study of *joinings* in ergodic theory. We recall the definition of this notion given for single transformations. Suppose that two dynamical systems, $(X, \mathcal{B}, \mu, \sigma)$ and $(Y, \mathcal{A}, \nu, \tau)$, are given. Then a joining is a measure λ on $(X \times Y, \mathcal{B} \times \mathcal{A})$ such that (i) λ is invariant with respect to $\sigma \times \tau$, and (ii) the projections of λ onto the X and Y coordinates are μ and ν, respectively. The theory of joinings is well developed in ergodic theory and topological dynamics and contains many impressive results. We refer to [Gla03, dlR06, Rud90] where the reader can find further references.

We finish this section by formulating a result that was proved in [AJL16].

Suppose a positive operator R, acting on measurable function over $(X, \mathcal{B}, \mu)$, has the properties

$$Rh = h, \qquad \mu R = \mu, \tag{4.26}$$

where h is a harmonic function for P and μ is a probability R-invariant measure.

Theorem 4.29 *Let R be a positive operator satisfying (4.26). Suppose*

$$(\Omega, \mathcal{B}_\infty) = \prod_0^\infty (X, \mathcal{B})$$

is the infinite product space. Then, on $(\Omega, \mathcal{B}_\infty)$, there exists a unique probability measure $\mathbb{P}$, defined on cylinder functions $f_0 \times f_1 \times \cdots \times f_n$ $(n \in \mathbb{N}, f \in \mathcal{F}(X, \mathcal{B}))$, as follows

$$\int_\Omega f_0 \times f_1 \times \cdots \times f_n \, d\mathbb{P} = \int_X f_0 R(f_1 R(\cdots R(f_{n-1} R(f_n h)) \cdots)) \, d\mu.$$

If $\{\pi_i \mid i = 0, 1, \ldots\}$ denotes the coordinate random functions, then the following Markov property holds

$$\mathbb{E}_\mathbb{P}(f \circ \pi_{i+1} \mid \pi_i^{-1}(x)) = R(f)(x)$$

for all i, $x \in X$, and $f \in \mathcal{F}(X, \mathcal{B})$.

Chapter 5
Transfer Operators on L^1 and L^2

Abstract Given a *transfer operator* (R, σ), it is of interest to find the *measures* μ such that both R and σ induce operators in the corresponding L^p spaces, i.e., in $L^p(X, \mathcal{B}, \mu)$. We turn to this below, but our main concern are the cases $p = 1$, $p = 2$, and $p = \infty$. When R is realized as an operator in $L^2(X, \mathcal{B}, \mu)$, for a suitable choice of μ, then it is natural to ask for the adjoint operator R^* where "adjoint" is defined with respect to the $L^2(\mu)$-inner product.

Keywords Induce operators · Universal Hilbert space · Infinite-dimensional Perron-Frobenius theorem

We turn to this question in Sects. 5.2 and 5.3 below. Our operator theoretic results for these L^2-spaces will be used in Chap. 8 below where we introduce a certain *universal Hilbert space* $\mathcal{H}(X)$, or rather $\mathcal{H}(X, \mathcal{B})$. Indeed, when a transfer operator (R, σ) is given, we show that there is then a naturally induced *isometry* in the universal Hilbert space $\mathcal{H}(X)$, which we show offers a number of applications and results which may be considered to be an infinite-dimensional Perron-Frobenius theory. Our study of transfer operators in L^2-spaces is motivated by [AJL16, Jor01].

Readers coming from other but related areas, may find the following papers useful for background, [BSV15, FMCB$^+$16, Mat17, Sil13, SW17, Sze17]. The following books are focused on basic properties of endomorphisms and transfer operators: [Bal00, CFS82, PU10].

5.1 Properties of Transfer Operators Acting on L^1 and L^2

In this section, we will keep the following settings. Let $(X, \mathcal{B}, \lambda)$ be a standard measure space, and let $\lambda \in M(X)$ be a Borel measure on $\mathcal{B}$. Suppose that σ is a non-singular surjective endomorphism on $(X, \mathcal{B}, \lambda)$.

© Springer International Publishing AG, part of Springer Nature 2018

S. Bezuglyi, P. E. T. Jorgensen, *Transfer Operators, Endomorphisms, and Measurable Partitions*, Lecture Notes in Mathematics 2217, https://doi.org/10.1007/978-3-319-92417-5_5

We will consider transfer operators (R, σ) defined on the space of Borel functions $\mathcal{F}(X, \mathcal{B})$. It will be assumed that the function $R(\mathbf{1})$ is either in $L^1(\lambda)$, or in $L^2(\lambda)$, depending on the context. We will point out explicitly the cases when R is assumed to be normalized.

Lemma 5.1 *Let $(X, \mathcal{B})$ be a standard Borel space, and $\sigma \in End(X, \mathcal{B})$. Set*

$$S(f) = f \circ \sigma, \quad f \in \mathcal{F}(X, \mathcal{B}).$$

For a measure μ on $(X, \mathcal{B})$, let the measure μS be defined by

$$\int_X f \, d(\mu S) = \int_X S(f) \, d\mu = \int_X f \circ \sigma \, d\mu. \tag{5.1}$$

Then $\mu S = \mu \circ \sigma^{-1}$.

Proof It follows from (5.1) that, for any $B \in \mathcal{B}$ and the characteristic function χ_B, we have

$$(\mu S)(B) = \int_X \chi_B \circ \sigma \, d\mu$$
$$= \int_X \chi_{\sigma^{-1}(B)} \, d\mu$$
$$= (\mu \circ \sigma^{-1})(B),$$

and the result follows.   □

Lemma 5.2 *Let (R, σ) be a transfer operator in $\mathcal{F}(X, \mathcal{B})$. Let λ be a Borel measure on $(X, \mathcal{B})$. Then (R, σ) induces a transfer operator in the space $L^p(\lambda)$ if and only if $\lambda R \ll \lambda$ and $\lambda \circ \sigma^{-1} \sim \lambda$.*

This observation is obvious and explains why we will work with the σ-quasi-invariant measures λ which belong to the set $\mathcal{L}(R)$.

Assume that $\lambda \in \mathcal{L}(R)$ denote the Radon-Nikodym derivative of λR with respect to λ by

$$W_\lambda(x) = W(x) := \frac{d(\lambda R)}{d\lambda}(x).$$

Since R is integrable, we have $W_\lambda \in L^1(\lambda)$, and the following useful equality holds (see (4.15)):

$$\int_X R(\mathbf{1}) \, d\lambda = \int_X W \, d\lambda, \quad \lambda \in \mathcal{L}(R).$$

Lemma 5.3 *In the above notation, the function $R(\mathbf{1})$ is represented as follows:*

$$R(\mathbf{1})(x) = \frac{(Wd\lambda) \circ \sigma^{-1}}{d\lambda}(x) = \frac{d(\lambda R) \circ \sigma^{-1}}{d\lambda}(x) \qquad (5.2)$$

where λ is any measure from $\mathcal{L}(R)$.

   If R is a normalized transfer operator, $R(\mathbf{1}) = \mathbf{1}$, then

$$(\lambda R) \circ \sigma^{-1} = \lambda, \qquad \forall \lambda \in \mathcal{L}(R).$$

   Moreover, a transfer operator R is integrable with respect to λ if and only if $(\lambda R)(X) < \infty$.

Proof By the definition of the Radon-Nikodym derivative W, we have the relation

$$\int_X R(f) \, d\lambda = \int_X fW \, d\lambda \qquad (5.3)$$

which holds for any measurable function f. In particular, f can be any simple function. Substitute $f \circ \sigma$ instead of f in (5.3). Then

$$\int_X (f \circ \sigma)W \, d\lambda = \int_X R(f \circ \sigma) \, d\lambda$$

$$= \int_X f R(\mathbf{1}) \, d\lambda.$$

Since the last relation holds for every f, we have the equality of measures

$$(Wd\lambda) \circ \sigma^{-1}(x) = R(\mathbf{1})d\lambda(x),$$

that proves the first statement.

   The other two assertions follow immediately from Definition 4.5 and (5.2).   □

   In the next remark we collect several direct consequences of Lemma 5.3. Though these results can be easily proved, they contain some important properties of transfer operators that are used below.

Remark 5.4

(1) Equality (5.2) might be confusing because the left hand side of the relation

$$R(\mathbf{1})(x) = \frac{d(\lambda R) \circ \sigma^{-1}}{d\lambda}(x), \ \lambda \in \mathcal{L}(R),$$

does not contain the measure λ. But we should remember that, in the setting introduced above, $R(1)(x)$ is considered as a function in $L^1(\lambda)$, so that λ is involved implicitly. If we denote by θ_λ the Radon-Nikodym derivative for a non-singular endomorphism σ,

$$\theta_\lambda(x) = \frac{d\lambda \circ \sigma^{-1}}{d\lambda}(x),$$

then the function $R(1)$ can be written as follows

$$R(1)(x) = \frac{d(\lambda R) \circ \sigma^{-1}}{d\lambda R}(x)\frac{d(\lambda R)}{d\lambda}(x)$$
$$= \theta_{\lambda R}(x)W(x).$$

(2) Let σ be a non-singular endomorphism of $(X, \mathcal{B}, \lambda)$. It follows from Lemma 5.3 that a transfer operator (R, σ) on $L^1(\lambda)$ is strict, i.e., $(R1)(x) > 0$ λ-a.e., if and only if $W(x) > 0$ λ-a.e., and $\lambda \circ \sigma^{-1} \sim \lambda$. Moreover, it is seen from (5.2) that we can prove the following result.

Lemma 5.5 *Let σ be a non-singular endomorphism of $(X, \mathcal{B}, \lambda)$. Then the following properties are equivalent:*

i)

$$R(1)(x) > 0 \ for \ \lambda\text{-}a.e. \ x;$$

ii)

$$W(x) > 0 \ for \ \lambda\text{-}a.e. \ x;$$

iii)

$$\lambda R \sim \lambda;$$

iv)

$$\theta_\lambda(x) = R(W^{-1}).$$

Proof We prove the equivalence of ii) and iv) only and leave other assertions to the reader. For this, we check

$$\int_X f\theta_\lambda \, d\lambda = \int_X (f \circ \sigma) \, d\lambda$$
$$= \int_X \frac{1}{W}W(f \circ \sigma) \, d\lambda$$

$$= \int_X R\left(\frac{1}{W}(f \circ \sigma)\right) d\lambda$$

$$= \int_X R\left(\frac{1}{W}\right) f \, d\lambda.$$

Since f is any function, we have the result. □

(3) If $R(\mathbf{1}) = \mathbf{1}$, then we obtain from the statements of Lemma 5.5 that

$$\theta_{\lambda R} = \frac{1}{W}, \qquad R(\theta_{\lambda R}) = \theta_\lambda.$$

Indeed, to see these, we find

$$\theta_{\lambda R} = \frac{d(\lambda R) \circ \sigma^{-1}}{d(\lambda R)} = \frac{d(\lambda R) \circ \sigma^{-1}}{d\lambda} \frac{d\lambda}{d(\lambda R)} = R(\mathbf{1})\frac{1}{W}. \tag{5.4}$$

(4) We notice that if (R, σ) is a strict transfer operator acting on the space of measurable functions over $(X, \mathcal{B}, \lambda)$ with $\lambda \in \mathcal{L}(R)$, then σ is non-singular with respect to λR. This fact follows from (5.4).

In other words, one has the properties

$$(\lambda R) \circ \sigma^{-1} \ll \lambda R \ll \lambda.$$

(5) Another corollary of relation (5.2) is formulated as follows: for any two measures $\lambda, \lambda' \in \mathcal{L}(R)$, we have

$$\frac{W_\lambda}{W_{\lambda'}} = \frac{\theta_{\lambda'R}}{\theta_{\lambda R}}.$$

In ergodic theory, it is extremely important to understand how properties of a transformation depend on a measure. More precisely, suppose a transformation T acts on a measure space $(X, \mathcal{B}, \mu)$. What can be said about dynamical properties of T if μ is replaced by an equivalent measure ν? We discuss here this question in the context of transfer operators.

Lemma 5.6 *Let (R, σ) be a transfer operator and $\lambda \in \mathcal{L}(R)$. Suppose that a Borel measure λ_1 is equivalent to λ, that is there exists a positive measurable function $\varphi(x)$ such that $d\lambda_1(x) = \varphi(x)d\lambda(x)$. Then $\lambda_1 \in \mathcal{L}(R)$, and the Radon-Nikodym derivative $W_1 = \dfrac{d\lambda_1 R}{d\lambda_1}$ is σ-cohomologous to W:*

$$W_1(x) = \varphi(\sigma(x))W(x)\varphi(x)^{-1}.$$

Proof The proof is based on the direct calculation, the definition of the Radon-Nikodym derivative for R, and the pull-out property of R. We note that because

$\lambda \sim \lambda_1$, then φ is positive a.e., so that W_1 is defined a.e. Let f be any measurable function, then we compute

$$\int_X R(f) \, d\lambda_1 = \int_X R(f)\varphi \, d\lambda$$

$$= \int_X R((\varphi \circ \sigma)f) \, d\lambda$$

$$= \int_X (\varphi \circ \sigma)f \, d(\lambda R)$$

$$= \int_X (\varphi \circ \sigma)f W \, d\lambda$$

$$= \int_X f(\varphi \circ \sigma)W\varphi^{-1} \, d\lambda_1$$

Thus, we proved that $(\lambda_1 R)(f) = (\varphi \circ \sigma)W\varphi^{-1}\lambda_1(f)$. Hence,

$$\frac{d\lambda_1 R}{d\lambda_1} = (\varphi \circ \sigma)W\varphi^{-1}.$$

$\square$

Remark 5.7 We observe that one can directly check the validity of the equality for the measure λ_1

$$R1 = \frac{[(\varphi \circ \sigma)W\varphi^{-1}d\lambda_1] \circ \sigma^{-1}}{d\lambda_1}.$$

This confirms the conclusion of Lemma 5.3.

Corollary 5.8 *Let (R, σ) be a transfer operator such that $R1 \in L^1(\lambda)$ for a Borel measure λ. Then, for $\lambda \in \mathcal{L}(R)$, the Radon-Nikodym derivative $W = \dfrac{d\lambda R}{d\lambda}$ is a coboundary with respect to σ if and only if there exists a measure λ_1 such that $\lambda_1 \sim \lambda$ and $\lambda_1 R = \lambda_1$.*

Proof Suppose that W is a coboundary, that is there exists a measurable function $q(x)$ such that $W = (q \circ \sigma)q^{-1}$. Take a new measure λ_1 defined by $d\lambda_1 = qd\lambda$. Then λ_1 is equivalent to λ and, for any integrable function f, we compute

$$\lambda_1(Rf) = \int_X f \, d(\lambda_1 R)$$

$$= \int_X R(f)q \, d\lambda$$

$$= \int_X R(f(q \circ \sigma)) \, d\lambda$$

$$= \int_X f(q \circ \sigma)) \, d(\lambda R)$$

$$= \int_X f(q \circ \sigma))(q \circ \sigma)^{-1} q \, d\lambda \qquad \text{(because } W = (q \circ \sigma)q^{-1})$$

$$= \int_X f \, d\lambda_1$$

$$= \lambda_1(f).$$

Hence λ_1 is R-invariant.

Conversely, suppose that a measure λ_1 is R-invariant and $d\lambda_1 = \varphi d\lambda$. By invariance with respect to R, we have

$$\int_X f \, d\lambda_1 = \int_X R(f) \, d\lambda_1$$

$$= \int_X R(f)\varphi \, d\lambda$$

$$= \int_X R(f(\varphi \circ \sigma)) \, d\lambda$$

$$= \int_X f(\varphi \circ \sigma)W \, d\lambda$$

$$= \int_X f(\varphi \circ \sigma)W\varphi^{-1} \, d\lambda_1$$

Since f is any function, we conclude that W is a σ-coboundary:

$$W = (\varphi \circ \sigma)^{-1}\varphi.$$

□

Theorem 5.9 *Let (R, σ) be a transfer operator acting in the space of Borel functions over $(X, \mathcal{B})$.*

(1) Suppose that R is such that the action $\lambda \mapsto \lambda R$ is well defined on the set of all measures $M(X, \mathcal{B})$. Then the partition of $M(X, \mathcal{B})$ into subsets $[\lambda] := \{\lambda' \in M(X, \mathcal{B}) : \lambda' \sim \lambda\}$, consisting of equivalent measures, is invariant with respect to the action of R. Thus, the transfer operator R sends equivalent measures to equivalent ones. More generally, if $\lambda_1 \ll \lambda$, then $\lambda_1 R \ll \lambda R$ and

$$\frac{d(\lambda_1 R)}{d(\lambda R)} = \varphi \circ \sigma$$

where $d\lambda_1 = \varphi d\lambda$.

(2) If $\lambda_1, \lambda_2 \in \mathcal{L}(R)$, then $\lambda_1 + \lambda_2 \in \mathcal{L}(R)$. Moreover,

$$W := \frac{d(\lambda_1 + \lambda_2)R}{d(\lambda_1 + \lambda_2)} = W_1 \frac{d\lambda_1}{d(\lambda_1 + \lambda_2)} + W_2 \frac{d\lambda_2}{d(\lambda_1 + \lambda_2)}$$

where W_i is the Radon-Nikodym derivative of R_i defined in (5.3), $i = 1, 2$.

Proof

(1) The first part of the statement is obvious, see Lemma 5.6 . To prove the other
statements in (1), it suffices to check the fact that $\lambda_1 \ll \lambda$ implies $\lambda_1 R \ll \lambda R$.
Since $d\lambda_1 = \varphi d\lambda$, we have

$$\int_X f \, d(\lambda_1 R) = \int_X R(f) \, d\lambda_1 = \int_X R(f) \varphi \, d\lambda$$

$$= \int_X R[(\varphi \circ \sigma) f] \, d\lambda = \int_X (\varphi \circ \sigma) f \, d(\lambda R).$$

Hence, we get

$$\varphi \circ \sigma = \frac{d(\lambda_1 R)}{d(\lambda R)}.$$

(2) We compute the Radon-Nikodym derivative of $(\lambda_1 + \lambda_2)R$ with respect to $(\lambda_1 + \lambda_2)$ as follows:

$$\int_X fW \, d(\lambda_1 + \lambda_2) = \int_X R(f) \, d(\lambda_1 + \lambda_2)$$

$$= \int_X f \, d(\lambda_1 R) + \int_X f \, d(\lambda_2 R)$$

$$= \int_X fW_1 \, d\lambda_1 + \int_X fW_2 \, d\lambda_2$$

$$= \int_X f \, [W_1 d\lambda_1 + W_2 d\lambda_2].$$

Thus, $Wd(\lambda_1 + \lambda_2) = W_1 d\lambda_1 + W_2 d\lambda_2$, and we are done.

□

Remark 5.10 Suppose that $(X, \mathcal{B}, \mu, \sigma)$ is a measurable dynamical system where σ
is a measurable transformation of X. How do properties of σ depend on the measure
μ? Can μ be replaced by an equivalent measure? These questions are well known
in ergodic theory, and many dynamical properties of σ do not depend on a choice
of a measure in the class $[\mu]$. In particular, this happens in the orbit equivalence
theory. The importance of Lemma 5.6 and Theorem 5.9 consists of explicit formulas
relating Radon-Nikodym derivatives of R to equivalent measures.

Remark 5.11

(1) It follows from Theorem 5.9 any transfer operator acts not only on individual measures from $\mathcal{L}(R)$ but also it acts on the set of classes of equivalent measures: $R[\lambda] = [\lambda R]$.

(2) We point out several formulas that relate the Radon-Nikodym derivatives for R and σ. They are based on Lemma 5.3 and Remark 5.4. It is assumed that a transfer operator (R, σ) is defined on $(X, \mathcal{B}, \lambda)$ where λ is a measure from $\mathcal{L}(R)$.

(a) If $\lambda R = \lambda$ (that is $W = 1$ a.e.), then

$$R(1)(x) = \frac{d\lambda \circ \sigma^{-1}}{d\lambda}(x) = \theta_\lambda(x). \qquad (5.5)$$

Let $Fix(R) := \{\lambda \in \mathcal{L}(R) : \lambda R = \lambda\}$ be the set of R-invariant measures. The above formula means that the function $(x, \lambda) \mapsto \theta_\lambda(x)$ does not depend on $\lambda \in Fix(R)$. As a confirmation of this observation, one can show directly that for $\lambda_1, \lambda_2 \in Fix(R)$ the condition $\theta_{\lambda_1+\lambda_2}(x) = R(1)(x)$ holds.

(b) If $\lambda R = \lambda$, then relation (5.5) implies that

$$R(1) = 1 \iff \theta_\lambda = 1.$$

Therefore, if $\lambda \in Fix(R)$, then R is normalized if and only if λ is σ-invariant.

(c) Let λ, λ_1 be two equivalent measures from $\mathcal{L}(R)$ where (R, σ) is a transfer operator. Let the function $\xi(x) > 0$, be defined by the relation $d\lambda_1(x) = \xi(x)d\lambda(x)$. Then we can show that

$$\frac{d(\lambda_1 R)\sigma^{-1}}{d\lambda_1}(x) = R(\xi \circ \sigma)(x)\frac{d\lambda}{d\lambda_1}(x)$$

5.2   The Adjoint Operator for a Transfer Operator

We recall briefly the notion of a symmetric pair of linear operators in a Hilbert space.

Suppose that $\mathcal{H}_1$ and $\mathcal{H}_2$ are Hilbert spaces and A and B are operators with dense domains $Dom(A) \subset \mathcal{H}_1$ and $Dom(B) \subset \mathcal{H}_2$ such that $A : Dom(A) \to \mathcal{H}_2$ and $B : Dom(B) \to \mathcal{H}_1$. It is said that $(A; B)$ is a *symmetric pair* if

$$\langle Ax, \ y\rangle_{\mathcal{H}_2} = \langle x, \ By\rangle_{\mathcal{H}_1},$$

where $x \in Dom(A), y \in Dom(B)$. In other words, $(A; B)$ is a symmetric pair if and only if $A \subset B^*$ and $B \subset A^*$.

If $(A; B)$ is a symmetric pair, then the operators A and B are closable. Moreover, one can prove that

1. $A^*\overline{A}$ is densely defined and self-adjoint with $Dom(A^*\overline{A}) \subset Dom(\overline{A}) \subset \mathcal{H}_1$,
2. $B^*\overline{B}$ is densely defined and self-adjoint with $Dom(B^*\overline{B}) \subset Dom(\overline{B}) \subset \mathcal{H}_2$.

Therefore, without loss of generality, we can assume that A and B are closed operators, and we can work with self-adjoint operators A^*A and B^*B.

We will discuss below transfer operators (R, σ) defined on $L^2(\lambda)$ where λ is a fixed σ-quasi-invariant measure. It turns out that one can explicitly describe various properties of R and the adjoint operator R^*.

Theorem 5.12 *Let (R, σ) be a transfer operator considered on the space $(X, \mathcal{B}, \lambda)$. Suppose that $R(1) \in L^1(\lambda) \cap L^2(\lambda)$ and $\lambda \circ R \ll \lambda$. Then R is a densely defined linear operator in the Hilbert space $\mathcal{H} = L^2(\lambda)$ whose adjoint operator R^* is determined by the formula*

$$R^*(f) = W(f \circ \sigma), \quad f \in \text{Dom}(R^*).$$

In particular, $W = R^*(1)$, and $W \in L^2(\lambda)$.

(We recall that the Radon-Nikodym derivative $W = W_\lambda$ is considered as a measurable function, and the displayed formula in Theorem 5.12 means the multiplication of W and $f \circ \sigma$. This remark remains true for other statements of this section, see e.g. Corollary 5.15.)

Proof We take simple functions f and g such that $f, g \in \mathcal{S}^2(\lambda)$. Then $(f \circ \sigma)g$ is also a simple function. Since $R(1) \in L^1(\lambda)$, we conclude that $R((f \circ \sigma)g) \in L^1(\lambda)$ according to Lemma 4.7.

It follows from (5.3) and (3.1) that the following equalities hold:

$$\int_X (f \circ \sigma)gW \, d\lambda = \int_X R((f \circ \sigma)g) \, d\lambda = \int_X fR(g) \, d\lambda.$$

All integrals in these formulas are well defined. Indeed, we use that $f, g \in \mathcal{S}^2(\lambda)$ and $R(1) \in L^2(\lambda)$ to conclude that $f, R(g)$ are in $L^2(\lambda)$. Hence, we have

$$\int_X fR(g) \, d\lambda = \langle R(g), f \rangle_{L^2(\lambda)}. \tag{5.6}$$

We recall that λR is a finite measure, and the functions $f \circ \sigma$, and g are simple. Therefore, the integrals

$$\int_X (f \circ \sigma)gW \, d\lambda = \int_X (f \circ \sigma)g \, d(\lambda R)$$

are finite. Moreover,

$$\int_X (f \circ \sigma) g W \, d\lambda = \langle g, W(f \circ \sigma) \rangle_{L^2(\lambda)}. \tag{5.7}$$

Thus, we have proved that, for any function $g \in \mathcal{S}^2(\lambda)$,

$$\langle R(g), f \rangle_{L^2(\lambda)} = \langle g, W(f \circ \sigma) \rangle_{L^2(\lambda)}.$$

This relation means that the adjoint operator R^* is defined for every $f \in \mathcal{S}^2(\lambda)$ and

$$R^*(f) = W(f \circ \sigma). \tag{5.8}$$

□

It turns out that when a transfer operator (R, σ) is considered as an operator in the space $L^2(\lambda)$ with $\lambda \in \mathcal{L}(R)$, then this operator can be realized explicitly, see Example 1.4.

Theorem 5.13 *Suppose R is a transfer operator acting in $L^2(\lambda)$ such that $d(\lambda R) = W d\lambda$. Then, for any $f \in L^2(\lambda)$,*

$$R(f)(x) = \frac{(fWd\lambda) \circ \sigma^{-1}}{d\lambda}(x), \qquad \lambda\text{-a.e.}. \tag{5.9}$$

Proof We first note that formula (5.9) defines a transfer operator that can be checked directly. Next, as it follows from (5.12), a function $R(f) \in L^2$ which satisfies the equation

$$\int_X g R(f) \, d\lambda = \int_X (g \circ \sigma) f W \, d\lambda, \qquad \forall g \in L^2(\lambda),$$

is uniquely determined by this relation. Then, the right hand side is represented as

$$\int_X (g \circ \sigma) f W \, d\lambda = \int_X g \, \frac{(fWd\lambda) \circ \sigma^{-1}}{d\lambda} \, d\lambda,$$

and this equality proves (5.9).   □

If (R, σ) is a transfer operator acting in $\mathcal{F}(X, \mathcal{B})$ and a measures $\lambda \in M(X)$ is such that $\lambda R \ll \lambda$, then the realization of R in $L^{(\lambda)}$, given in (5.9), is denoted by R_λ.

Proposition 5.14 *Let a transfer operator (R_λ, σ) be defined in $L^2(\lambda)$ and suppose that $R_\lambda(1) = 1$. Then, for any measure $\lambda' \sim \lambda$, the operator $R_{\lambda'} \in L^2(\lambda)$ has the property $R_{\lambda'} 1 = 1$.*

Proof Let $d\lambda' = \varphi d\lambda$. Then, as shown in Lemma 5.6, the corresponding Radon-Nikodym derivatives W_λ and $W_{\lambda'}$ are related by the formula

$$W_{\lambda'} = (\varphi \circ \sigma) W_\lambda \varphi^{-1}.$$

Based on the proof of Theorem 5.13, it suffices to show that, for any $g \in L^2(\lambda')$,

$$\int_X g \, d\lambda' = \int_X (g \circ \sigma) W_{\lambda'} \, d\lambda'.$$

We compute

$$\int_X (g \circ \sigma) W_{\lambda'} \, d\lambda' = \int_X (g \circ \sigma)(\varphi \circ \sigma) W_\lambda \varphi^{-1} \varphi \, d\lambda$$

$$= \int_X [(g\varphi) \circ \sigma] W_\lambda \, d\lambda$$

$$= \int_X R[(g\varphi) \circ \sigma] \, d\lambda$$

$$= \int_X g\varphi \, d\lambda$$

$$= \int_X g \, d\lambda'.$$

This means that $R_{\lambda'}(\mathbf{1}) = \mathbf{1}$. $\qquad\square$

The following corollary is basically deduced from Theorem 5.12.

Corollary 5.15 *Let R be a transfer operator in $L^2(\lambda)$ such that $d(\lambda R) = W d\lambda$ and $R(\mathbf{1}) \in L^1(\lambda) \cap L^2(\lambda)$.*

(1) *The domain of R^* contains the dense set $S^2(\lambda)$. The transfer operator R is closable in $L^2(\lambda)$.*
(2) *The following formulas hold:*

$$R^*(f) = (f \circ \sigma) R^*(\mathbf{1}),$$

$$(RR^*)(f) = f(RR^*)(\mathbf{1}) = R(W) f.$$

This means that RR^* is a multiplication operator.
(3) *$R^*(f) \in L^1(\lambda)$ for any simple function f.*
(4) *R^* is an isometry if and only if $R(W) = \mathbf{1}$.*
(5) *For every $n \in \mathbb{N}$, the operator $(R^*)^n$ is defined on $L^2(\lambda)$ by the formula*

$$(R^*)^n f = (f \circ \sigma^n) W(W \circ \sigma) \cdots (W \circ \sigma^{n-1}).$$

Proof The proofs of most statements are rather obvious so that they can be left for the reader. We show here the proof of (4) only. Let $f, g \in L^2(\lambda)$, then we have

$$
\begin{aligned}
\langle R^* f, R^* g \rangle &= \int_X W(f \circ \sigma) W(g \circ \sigma) \, d\lambda \\
&= \int_X W((fg) \circ \sigma) \, d(\lambda R) \\
&= \int_X R[W((fg) \circ \sigma)] \, d\lambda \\
&= \int_X (fg) R(W) \, d\lambda R
\end{aligned}
$$

Hence, $\langle R^* f, R^* g \rangle = \langle f, g \rangle$ if and only if $R(W) = 1$.   □

Proposition 5.16 *Let the conditions of Theorem 5.12 hold. Then the domain of R^n, considered as an operator in $L^2(\lambda)$, contains functions $f \in S^2(\lambda)$ for any $n \geq 2$.*

Proof Suppose $n = 2$. Then, for $f \in S^2(\lambda)$, we see that

$$
\int_X R^2(f) \, d\lambda = \int_X R(R(f)) \, d\lambda = \int_X R(f) W \, d\lambda.
$$

Since $R(f)$ and W are in $L^2(\lambda)$ (see Theorem 5.12), we conclude that

$$
\int_X R^2(f) \, d\lambda = \langle R(f), W \rangle_{L^2(\lambda)} < \infty.
$$

To prove the statement for any natural $n > 2$, we use induction.   □

Proposition 5.16 is important for consideration powers of R because it states that R^n has a dense domain in $L^2(\lambda)$ for every n.

Proposition 5.17 *Let (R, σ) be a transfer operator acting in the $L^2(\lambda)$-space of measurable functions over a measure space $(X, \mathcal{B}, \lambda)$. Assume that there is a nontrivial harmonic function for (R, σ). Suppose also that $\lambda R \ll \lambda$ and let $W d\lambda = d(\lambda R)$. Then we have*

$$
\|R(W)\|_{L^\infty(\lambda)} \leq 1 \implies R(W) = 1 \ a.e.
$$

The converse is obviously true.

Proof We use Corollary 5.15 to prove the following relation

$$
\|RR^*\|^2_{L^2(\lambda)} = \|R(W)\|^2_{L^\infty(\lambda)}
$$

where $\|T\|_{L^2(\lambda)}$ denotes the operator norm of T when $T : L^2(\lambda) \to L^2(\lambda)$ is a bounded operator. It follows then that $\|R\|_{L^2(\lambda)} = \|R^*\|_{L^2(\lambda)} \le 1$, i.e. R and R^* are contractions in $L^2(\lambda)$. We notice that, in this case,

$$Rh = h \iff R^*h = h.$$

(For this, one can show that $\|R^*h - h\|_{L^2(\lambda)} \le 0 \iff R^*h = h$). Therefore, we use Theorem 5.12 to deduce that

$$Rh = h \implies h = W(h \circ \sigma), \tag{5.10}$$

that is W is a σ-coboundary. When we apply R to the right hand side of (5.10), then we obtain $h = R(W)h$ a.e., hence $R(W) = 1$ a.e.   □

Remark 5.18 We observe that, due to the Schwarz inequality, the following relation is true.

$$|R(f)| \le \sqrt{R(|f|^2)}\sqrt{(R(1))}.$$

Here we assume that R is integrable and $R(f) \in L^2(\lambda)$. In particular, this holds for simple functions.

More generally, we have that, for any $k \in \mathbb{N}$,

$$|R(f)| \le \mathbb{R}(|f|^{2^k})^{2^{-k}} R(1)^{\sum_{i=1}^k 2^{-i}}.$$

5.3   More Relations Between R and σ

Let $(X, \mathcal{B}, \lambda)$ be a measure space with a surjective endomorphism σ. We recall our assumption about the endomorphism σ: it is forward and backward quasi-invariant, i.e., $\lambda(A) = 0$ if and only if $\lambda(\sigma^{-1}(A)) = 0$ if and only if $\lambda(\sigma(A)) = 0$. Let (R, σ) be a transfer operator such that $\lambda \circ R \ll \lambda$. We will focus here on the study of relations between the transfer operator (R, σ) and the endomorphism σ.

For R and σ, we recall the definitions of the Radon-Nikodym derivatives (see Remark 2.3)

$$W = \frac{d\lambda R}{d\lambda} \quad \text{and} \quad \omega_\lambda = \frac{d\lambda \circ \sigma}{d\lambda}.$$

Proposition 5.19 *In the above notation, we have*

$$R(\omega_\lambda R) = W.$$

Proof It follows from the definition of the Radon-Nikodym derivative ω_λ that, for any $f \in L^1(\lambda)$, one has

$$\int_X f \, d\lambda = \int_X (f \circ \sigma)\omega_\lambda \, d\lambda$$

where ω_λ is a uniquely determined function which is measurable with respect to $\sigma^{-1}(\mathcal{B})$. We apply this equality to the measure λR and obtain the following sequence of equalities:

$$\int_X (f \circ \sigma)\omega_{\lambda R} \, d(\lambda R) = \int_X f \, d(\lambda R)$$

$$\updownarrow$$

$$\int_X R[(f \circ \sigma)\omega_{\lambda R}] \, d\lambda = \int_X R(f) \, d\lambda$$

$$\updownarrow$$

$$\int_X f R(\omega_{\lambda R}) \, d\lambda = \int_X f W \, d\lambda.$$

Since f is an arbitrary function from $L^1(\lambda R)$, we obtain the desired result. □

We recall that a measure $\lambda \in \mathcal{L}(R)$ is called *invariant* with respect to R if $\lambda R = \lambda$, i.e., $W(x) = 1$ for λ-a.e. x.

Let $\varphi(x)$ be a Borel non-negative function. Given $\lambda \in \mathcal{L}(R)$, we define the measure μ:

$$d\mu(x) = \varphi(x)d\lambda(x) \qquad (5.11)$$

Then, for any measurable f,

$$\int_X f \, d\mu = \int_X f\varphi \, d\lambda.$$

Theorem 5.20 *Given a transfer operator (R, σ), suppose that there exists a Borel measure λ on $(X, \mathcal{B})$ such that $R(1) \in L^1(\lambda)$ and $\lambda R = \lambda$. Then a Borel measure μ defined by its λ-density $\varphi(x)$ as in (5.11) is R-invariant if and only if $\varphi \circ \sigma = \varphi$.*

Proof The proof follows from the following chain of equalities. Let g be a measurable function, then

$$\int_X g \, d(\mu R) = \int_X R(g) \, d\mu$$

$$= \int_X R(g)\varphi \, d\lambda$$

$$= \int_X R[(\varphi \circ \sigma)g] \, d\lambda$$

$$= \int_X (\varphi \circ \sigma)g \, d(\lambda R)$$

$$= \int_X (\varphi \circ \sigma)g \, d\lambda$$

Hence, if $\mu \circ R = \mu$, we get from the above relations that

$$\int_X \varphi g \, d\lambda = \int_X (\varphi \circ \sigma)g \, d\lambda,$$

and $\varphi \circ \sigma = \varphi$. Conversely, if $\varphi \circ \sigma = \varphi$ holds, then

$$\int_X g \, d(\mu R) = \int_X g \, d\mu,$$

that is $\mu R = \mu$. $\square$

We can easily deduce from Theorem 5.20 (see the next statement) that if σ is ergodic on $(X, \mathcal{B}, \lambda)$, then any two R-invariant measures are proportional.

Corollary 5.21 *Let (R, σ) be a transfer operator on $(X, \mathcal{B}, \lambda)$ such that $\lambda R = \lambda$. Suppose that σ is an ergodic endomorphism with respect to the measure λ. Then a Borel measure $\mu \ll \lambda$ is R-invariant if and only if $\mu = c\lambda$ for some $c \in \mathbb{R}_+$.*

The following result clarifies the relationship between harmonic functions for a transfer operator R and σ-invariant measures μ.

Proposition 5.22 *Let (R, σ) be a transfer operator on $(X, \mathcal{B}, \lambda)$ such that $\lambda R = \lambda$. Let h be a non-negative Borel function. Then h is R-harmonic, $Rh = h$, if and only if the measure $d\mu(x) := h(x)d\lambda(x)$ is σ-invariant, i.e., $\mu \circ \sigma^{-1} = \mu$.*

Proof The proof follows from the following argument. Let g be an arbitrary Borel function. Then we deduce that the relation $Rh = h$ implies that $\mu \circ \sigma^{-1} = \mu$:

$$\int_X g \, d(\mu \circ \sigma^{-1}) = \int_X g \circ \sigma \, d\mu$$

$$= \int_X (g \circ \sigma)h \, d\lambda$$

$$= \int_X R[(g \circ \sigma)h] \, d\lambda, \qquad \text{(recall that } \lambda R = \lambda)$$

$$= \int_X g R(h) \, d\lambda$$

$$= \int_X g \, d\mu.$$

Conversely, if we assume that μ is σ-invariant, then we can show, in a similar way, that

$$\int_X gh \, d\lambda = \int_X g R(h) \, d\lambda$$

for arbitrary function g. Hence, h is harmonic for R.   □

We recall that if (R, σ) is a transfer operator and k is a positive Borel function, then one can define a new transfer operator (R_k, σ) where $R_k(f) = R(fk)k^{-1}$ (in fact, k can be non-negative but this generalization is inessential). Furthermore, this operator is normalized when k is R-harmonic. More details are in Sect. 3.5.

Lemma 5.23 *Let (R, σ) be a transfer operator and let k be a positive function. Suppose that $\lambda \in \mathcal{L}(R)$ and denote by W the corresponding Radon-Nikodym derivative, $Wd\lambda = d\lambda R$. Then the measure λ_k such that $d\lambda_k = kd\lambda$ is in $\mathcal{L}(R_k)$ and $W_k = W$ where $W_k d\lambda_k = d(\lambda_k R_k)$. In particular, if $\lambda R = \lambda$, then $\lambda_k = \lambda_k R_k$.*

Proof For any integrable function f, we get

$$\int_X f \, d(\lambda_k R_k) = \int_X R_k(f) \, d\lambda_k$$

$$= \int_X R(fk)k^{-1}k \, d\lambda$$

$$= \int_X fkW \, d\lambda$$

$$= \int_X fW \, d\lambda_k.$$

This proves that the Radon-Nikodym derivative W_k of measures $\lambda_k R_k$ and λ_k is W for any positive function k and any measure $\lambda \in \mathcal{L}(R)$.   □

Lemma 5.24 *Let (R, σ) be a transfer operator considered on $L^2(\lambda)$ where $\lambda \in \mathcal{L}(R)$. Let W denote the Radon-Nikodym derivative $\dfrac{d(\lambda R)}{d\lambda}$. Suppose that R possesses a harmonic function h, and we set $d\lambda_h = hd\lambda$. Then the operator*

$$V : f \mapsto W^{1/2}(f \circ \sigma)$$

is an isometry in $L^2(\lambda_h)$.

Proof We verify by direct calculations that $\|Vf\|_{L^2(\lambda_h)} = \|f\|_{L^2(\lambda_h)}$:

$$\int_X |(Vf)|^2 \, d\lambda_h = \int_X W(|f|^2 \circ \sigma)h \, d\lambda$$

$$= \int_X (|f|^2 \circ \sigma)h \, d(\lambda R)$$

$$= \int_X R[(|f|^2 \circ \sigma)h] \, d\lambda$$

$$= \int_X |f|^2 R(h) \, d\lambda$$

$$= \int_X |f|^2 h \, d\lambda$$

$$= \int_X |f|^2 \, d\lambda_h.$$

This proves the result.   □

Chapter 6
Actions of Transfer Operators on the Set of Borel Probability Measures

Abstract Let (R, σ) be a transfer operator defined on the space of Borel functions $\mathcal{F}(X, \mathcal{B})$. The main theme of this chapter is the study of a dual action of R on the set of probability measures $M_1 = M_1(X, \mathcal{B})$. As a matter of fact, a big part of our results in this chapter remains true for any sigma-finite measure on $(X, \mathcal{B})$, but we prefer to work with probability measures. The justification of this approach is contained in the results of Chap. 5 where we showed that the replacement of a measure by a probability measure does not affect the properties of R described in terms of measures. Our main assumption for this chapter is that the transfer operators R are normalized, that is $R(1) = 1$. In other chapters, we also used this assumption to prove some results.

Keywords Probability measures · Normalized transfer operator · R-invariant measures

We recall that there are classes of transfer operators R for which the natural action $\lambda \mapsto \lambda R$ on the set of measures is well defined. For instance, this is true for order continuous transfer operators, and for transfer operators defined on locally compact Hausdorff space, see Sect. 4.1. The advantage of dealing with probability measures and normalized operators R is that, in this case, the measure $\lambda R \in M_1$ is defined for any measure $\lambda \in M_1$ and for any normalized operator R, see Proposition 4.10.

It follows from the above remark that, given a normalized transfer operator (R, σ), we can associate two maps defined on M_1. They are

$$t_R : \lambda \mapsto \lambda R, \qquad s_\sigma : \lambda \mapsto \lambda \circ \sigma^{-1}.$$

We call the maps t_R and s_σ *actions* of R and σ on M_1, respectively.

For a normalized transfer operator (R, σ), we can find out how these maps interact. We will show that the map t_R is one-to-one but not onto (this fact is proved below in Theorem 6.9). Thus, we get the following decreasing sequence of subsets:

$$M_1 \supset M_1 R \supset M_1 R^2 \supset \cdots.$$

We will use the notation $K_i(R) = M_1(X)R^i$. Our interest is mostly focused on the set $K_1(R)$ (or simply K_1 when R is fixed) because this set is crucial in our study of the action of R on measures.

For a transfer operator (R, σ), we also define the set of R-invariant measures and the set of σ-invariant measures, by setting

$$\mathrm{Fix}(R) := \{\lambda \in M_1 : \lambda R = \lambda\},$$

$$\mathrm{Fix}(\sigma) := \{\lambda \in M_1 : \lambda \circ \sigma^{-1} = \lambda\}.$$

We are interested in the following *question.* Under what conditions on a transfer operator (R, σ) is the set of invariant measures $\mathrm{Fix}(R)$ non-empty? A partial answer was given in Theorem 5.20.

Remark 6.1 We recall that, for a fixed transfer operator R, we dealt with the subset $\mathcal{L}(R)$ of the set of all measures M_1: by definition, $\lambda \in \mathcal{L}(R)$ if and only if $\lambda R \ll \lambda$. In particular, λR can be equivalent to λ. Theorem 5.9 asserts that when the set of all measures M_1 is partitioned into the classes of equivalent measures $[\lambda] := \{\mu : \mu \sim \lambda\}$, then the map t_R preserves the partition into these classes $[\lambda]$. The same holds for the sets $[\lambda]_\ll := \{\nu : \nu \ll \lambda\}$. These facts are obviously true for the action of s_σ. Thus, if $M_1(\sim)$ is the set of classes of equivalent measures, then t_R and s_σ induce the maps $t_R(\sim)$ and $s_\sigma(\sim)$, defined on $M_1(\sim)$.

These facts will be used in Chap. 8 in the construction of the universal Hilbert space.

The action s_σ of σ on the measure space M_1 is used to define the following two subsets naturally related to σ:

$$\mathcal{Q}_+(\sigma) := \{\lambda \in M_1 : \lambda \circ \sigma \ll \lambda\}$$

and

$$\mathcal{Q}_-(\sigma) := \{\lambda \in M_1 : \lambda \circ \sigma^{-1} \ll \lambda\}.$$

We begin with a simple observation about measures for powers of a transfer operator R.

Lemma 6.2 *Let R be a transfer operator acting on a functional space over* $(X, \mathcal{B})$, *and* $\mathcal{L}(R) = \{\lambda \in M_1 : \lambda R \ll \lambda\}$. *Then*

$$\mathcal{L}(R) \subset \mathcal{L}(R^2) \subset \cdots \mathcal{L}(R^n) \subset \mathcal{L}(R^{n+1}) \subset \cdots,$$

$$\mathcal{Q}_-(\sigma) \subset \mathcal{Q}_-(\sigma^2) \subset \cdots \subset \mathcal{Q}_-(\sigma^n) \subset \mathcal{Q}_-(\sigma^{n+1}) \cdots$$

Proof This fact follows immediately from Theorem 5.9 and the discussion in Remark 6.1. □

Lemma 6.3 *Suppose that* $\lambda \in \mathcal{L}(R)$ *and* $\mu \ll \lambda$. *Then* $\mu \in \mathcal{L}(R)$.

Proof We need to show that $\mu R \ll \mu$. Since $\lambda R \ll \lambda$ and $\mu \ll \lambda$, there exist measurable functions φ and W from $L^1(\lambda)$ such that

$$\varphi = \frac{d\mu}{d\lambda}, \quad W = \frac{d(\lambda R)}{d\lambda}.$$

Set

$$Q(x) = \begin{cases} ((\varphi \circ \sigma)W\varphi^{-1})(x), & \text{if } x \in A := \{x : \varphi(x) \neq 0\} \\ 0, & \text{if } x \in A^c := \{x : \varphi(x) = 0\} \end{cases}.$$

Take any measurable function f and compute

$$\int_X f \, d(\mu R) = \int_X R(f) \, d\mu$$

$$= \int_A R(f)\varphi \, d\lambda$$

$$= \int_A R(f(\varphi \circ \sigma)) \, d\lambda$$

$$= \int_A f(\varphi \circ \sigma) \, d(\lambda R)$$

$$= \int_A f(\varphi \circ \sigma)W \, d\lambda$$

$$= \int_A f(\varphi \circ \sigma)\varphi^{-1}W\varphi \, d\lambda$$

$$= \int_A f(\varphi \circ \sigma)W\varphi^{-1} \, d\mu$$

$$= \int_X fQ \, d\mu.$$

This proves that $\mu R \ll \mu$, and $Q = \dfrac{d(\mu R)}{d\mu}$.

□

In the following lemmas we study the relations between the maps t_R, s_σ and the sets $\mathcal{L}(R)$, $\mathcal{Q}_-(\sigma)$, $\mathcal{Q}_+(\sigma)$.

Lemma 6.4 *If* $\lambda \in \mathcal{Q}_-(\sigma)$, *then* $\lambda \ll \lambda \circ \sigma$.

Proof For any Borel set, one has $A \subset \sigma^{-1}(\sigma(A))$. Therefore, if $(\lambda \circ \sigma)(A) = 0$, then $(\lambda \circ \sigma^{-1})(\sigma(A)) = 0$, and then $\lambda(A) = 0$. □

Lemma 6.5 *If* (R, σ) *is a normalized transfer operator, then, for any measure* λ *and Borel set* A,

$$\lambda(\{x \in X : R(\chi_A)(x) > 0\}) = (\lambda \circ \sigma)(A).$$

This fact follows immediately from Lemma 4.7.

Theorem 6.6 *If* (R, σ) *is a normalized transfer operator, then*

$$\mathcal{L}(R) = \mathcal{Q}_+(\sigma).$$

Proof ($\subset$) Suppose that $\lambda R \ll \lambda$. We proved in Theorem 4.14 that, for any Borel set A,

$$(\lambda R)(A) = \int_{\sigma(A)} R(\chi_A)\, d\lambda. \tag{6.1}$$

If $\lambda(A) = 0$ implies that $(\lambda R)(A) = 0$, then, by (6.1) and Lemma 6.5, we conclude that $\lambda(\sigma(A)) = 0$.

($\supset$) Conversely, if $\lambda(A) = 0$ implies that $(\lambda \circ \sigma)(A) = 0$, then we again use (6.1) and obtain that $\int_{\sigma(A)} R(\chi_A)\, d\lambda = 0$, that is $(\lambda R)(A) = 0$. □

Lemma 6.7 *Let* (R, σ) *be a transfer operator with* $R(1) = 1$. *Then*

$$\lambda \in \mathcal{L}(R) \bigcap \mathcal{Q}_-(\sigma) \iff \lambda R \sim \lambda.$$

Proof We need to show only that $\lambda \ll \lambda R$. This is equivalent to the statement that $\lambda(A) > 0$ implies $(\lambda R)(A) > 0$. Since λ is a quasi-invariant measure with respect to σ and $A \subset \sigma^{-1}(\sigma(A))$, we see that $\lambda(\sigma(A)) > 0$, and therefore

$$(\lambda R)(A) = \int_X R(\chi_A)\, d\lambda > 0.$$

Conversely, suppose that $\lambda R \sim \lambda$. Then we have to show that $\lambda \circ \sigma^{-1} \ll \lambda$. The fact that $R1 = 1$ implies that $(\lambda R) \circ \sigma^{-1} = \lambda$ for any measure λ. Since the map s_σ preserves the partition of M_1 into the classes of equivalent measures, we obtain that $\lambda \circ \sigma^{-1} \ll \lambda$ (in fact we have that these measures are equivalent). This proves the statement. □

Lemma 6.8 *Let v be a measure from $\mathcal{L}(R)$. Then for $\lambda = vR$ we have the property*

$$\frac{d\lambda R}{d\lambda} \in \mathcal{F}(X, \sigma^{-1}(\mathcal{B})).$$

More generally, $\dfrac{d\lambda R}{d\lambda}$ is $\sigma^{-i}(\mathcal{B})$-measurable if $\lambda = vR^i$ and $v \in \mathcal{L}(R)$ and $i \in \mathbb{N}$.

Proof In order to prove the result, it suffices to note that due to Theorem 5.9

$$\frac{d\lambda R}{d\lambda} = \frac{dvR^2}{dvR} = \frac{dvR}{dv} \circ \sigma.$$

$\square$

In the following lemma, we collect a number of results that follow from the proved lemmas and definitions given in this section.

Theorem 6.9 *Let (R, σ) be a normalized transfer operator acting in the space of Borel functions $\mathcal{F}(X, \mathcal{B})$ such that the dual action $\lambda \mapsto \lambda R : M_1 \to K_1(R) = M_1 R$ is well defined. Then the following six statements hold:*

(1) A measure $\mu \in K_1(R)$ if and only if $(\mu \circ \sigma^{-1})R = \mu$.
(2) For any measure μ, the equation $\mu = \lambda R$ has a unique solution $\lambda = \mu \circ \sigma^{-1}$.
(3) The map t_R is one-to-one on M_1.
(4a) Two measures λ and λ' are mutually singular if and only if the measures λR and $\lambda' R$ are mutually singular.
(4b) If $\lambda \in K_1(R)$, then λ and $\lambda \circ \sigma^{-1}$ are mutually singular if and only if λ and λR are mutually singular.
(5)

$$Fix(R) \subset \bigcap_{i=0}^{\infty} M_1 R^i.$$

(6)

$$K_1(R) \bigcap Fix(\sigma) = Fix(R).$$

Proof

(1) We first recall that the condition $R(\mathbf{1}) = \mathbf{1}$ can be written in an equivalent form, namely,

$$(\lambda R) \circ \sigma^{-1} = \lambda, \quad \forall \lambda \in M_1.$$

Hence, if $\mu \in K_1(R)$, then $\mu = \lambda R$ for some $\lambda \in M_1$, and

$$[\mu \circ \sigma^{-1}]R = [(\lambda R) \circ \sigma^{-1}]R = \lambda R = \mu.$$

The converse is obvious.

(2) This fact follows immediately from statement (1).

(3) Suppose that $\lambda R = \lambda' R$. Then condition $R(1) = 1$ implies that

$$\lambda = (\lambda R) \circ \sigma^{-1} = (\lambda' R) \circ \sigma^{-1} = \lambda',$$

and statement (3) is proved.

(4) Suppose λ and λ' are mutually singular measures. Then there exists a set A such that $\lambda(A) = 1$ and $\lambda'(A) = 0$. Then

$$(\lambda R)(\sigma^{-1}(A)) = \int_X \chi_{\sigma^{-1}(A)} d(\lambda R) = \int_X \chi_A R(1) d\lambda = \lambda(A) = 1$$

and, similarly, we get that $(\lambda' R)(\sigma^{-1}(A)) = \lambda'(A) = 0$. To see that the converse is true, we observe that if λR and $\lambda' R$ are singular, then, applying s_σ to these measures, we obtain that λ and λ' are singular. This proves (4a)

To show that (4b) holds, we use (4a) and note that if λ and $\lambda \circ \sigma^{-1}$ are mutually singular, then, applying t_R to these measures, we get that λ and λR are mutually singular. To see that the converse holds we begin with mutually singular measures λ and λR and apply s_σ to them. Since R is normalized, the result follows.

(5) This statement is obvious.

(6) If $\lambda \in K_1(R) \cap \mathrm{Fix}(\sigma)$, then $\lambda = (\lambda \circ \sigma^{-1})R = \lambda R$. Conversely, if $\lambda = \lambda R$, then $\lambda \in K_1(R)$ by (5), hence $(\lambda \circ \sigma^{-1})R = \lambda R$. Since t_R is a one-to-one map, we conclude that $\lambda \circ \sigma^{-1} = \lambda$.

□

In the next lemma, we continue discussing relations between the maps t_R and s_σ for a transfer operator (R, σ).

Lemma 6.10 *Let (R, σ) be a transfer operator such that $R(1) = 1$. The following statements hold.*

(1) $s_\sigma t_R = id_{M_1}$ and $t_R s_\sigma = id_{K_1}$ where $K_1 = M_1 R$.

(2) If $\lambda \in K_1$, then

$$\lambda \circ \sigma^{-1} \ll \lambda \iff \lambda \ll \lambda R; \qquad (6.2)$$

$$\lambda \circ \sigma^{-1} = \lambda \iff \lambda = \lambda R; \qquad (6.3)$$

$$\lambda \circ \sigma^{-1} \gg \lambda \iff \lambda \gg \lambda R. \qquad (6.4)$$

(3) Let $T(\lambda) := t_R s_\sigma(\lambda)$. Then $T : M_1 \to K_1$ and $T^2 = T$. Moreover, if $\lambda_1 = T(\lambda)$, then

$$\lambda_1 \circ \sigma^{-1} = \lambda \circ \sigma^{-1}.$$

Proof

(1) The statement $s_\sigma t_R = \text{id}_{M_1}$ is a reformulation of the fact that $R(1) = 1$ (see, for example, Theorem 6.9 (1)). Let $\lambda \in K_1$, then it follows that $t_R s_\sigma = \text{id}_{K_1(R)}$.

(2) If $\lambda \circ \sigma^{-1} \ll \lambda$, then, because R possesses the "monotonicity" property ($\mu \ll \nu \implies \mu R \ll \nu R$), we obtain that $\lambda = (\lambda \circ \sigma^{-1}) R \ll \lambda R$. Conversely, suppose that $\lambda \ll \lambda R$. Then, applying σ^{-1} to this relation, we have $\lambda \circ \sigma^{-1} \ll (\lambda R) \circ \sigma^{-1} = \lambda$. This proves (6.2).

Relation (6.3) was proved in Theorem 6.9 (6).

To show that (6.4) holds, we again apply R to the both sides of $\lambda \circ \sigma^{-1} \gg \lambda$ and get that $\lambda R \ll \lambda$. The converse implication, i.e., $\lambda R \ll \lambda$ implies $\lambda \circ \sigma^{-1} \gg \lambda$, follows from the fact $R1 = 1$ and application of σ^{-1} to $\lambda R \ll \lambda$. Observe that this implication is true for any measure λ.

(3) The fact that $T^2 = T$ follows from the property $R1 = 1$ and the corresponding relation $s_\sigma t_R = \text{id}_{M_1}$.

Because $\lambda_1 = T(\lambda) = (\lambda \circ \sigma^{-1}) R$, then, taking into account that R is a normalized operator, we obtain

$$\lambda_1 \circ \sigma^{-1} = [(\lambda \circ \sigma^{-1}) R] \circ \sigma^{-1} = \lambda \circ \sigma^{-1}.$$

$\square$

Remark 6.11 We note that for any measure λ in $M(X)$, the following relation holds

$$\lambda R \ll \lambda \circ \sigma.$$

Indeed, this claim easily follows from Lemma 4.7 because the function $R(\chi_A)$ takes zero value on the complement of $\sigma(A)$.

Chapter 7
Wold's Theorem and Automorphic Factors of Endomorphisms

Abstract In this chapter, we discuss Wold's theorem stating the existence of a decomposition of any isometry operator of a Hilbert space in a unitary part and a unilateral shift.

Keywords Isometry · Decompositions · Wold · Stochastic · Orthogonal direct sum · Automorphic factors · Exact endomorphisms

7.1 Hilbert Space Decomposition Defined by an Isometry

The variant of Wold's theorem, we outline below, is a bit more geometric than the original result of Wold, which was, in fact, a decomposition theorem stated for stationary stochastic processes. The geometric variant is a result that applies to the wider context of any isometry in a Hilbert space. Some of the relevant references include [Wol48, Wol51, Wol54, HW70, BJ02].

Let $\mathcal{H}$ be a real Hilbert space, and let S be an isometry in $\mathcal{H}$. This means that $\|Sx\| = \|x\|$ for every $x \in \mathcal{H}$, or equivalently, $S^*S = I$ where I denotes the identity operator in $\mathcal{H}$. In general, S is not surjective.

It follows immediately that the operator $E_1 = SS^*$ is a projection. More generally, one can show that $E_n := S^n(S^*)^n$ is a projection. Indeed, we use n times the relation $S^*S = I$ and obtain

$$E_n^2 = S^n(S^*)^{n-1}(S^*S)S^{n-1}(S^*)^n$$

$$= S^n(S^*)^{n-1}S^{n-1}(S^*)^n = \cdots$$

$$= S^n(S^*)^n$$

$$= E_n.$$

Lemma 7.1 *The sequence of projections $\{E_n\}$ is decreasing*

$$I \geq E_1 \geq E_2 \geq \cdots ,$$

and each $E_n : \mathcal{H} \rightarrow S^n(\mathcal{H})$ is onto.

This result follows from the obvious relation:

$$\mathcal{H} \supset S(\mathcal{H}) \supset S^2(\mathcal{H}) \supset \cdots .$$

Let $R_S := \{Sx : x \in \mathcal{H}\}$ be the range of S. Consider the kernel of the adjoint operator

$$N_{S^*} := \{x \in \mathcal{H} : S^*x = 0\}.$$

Then one can see that

$$(N_{S^*})^\perp = R_S, \quad N_{S^*} = (R_S)^\perp. \tag{7.1}$$

More generally, if V is a bounded linear operator in $\mathcal{H}$, then

$$ker(V^*) = \mathcal{H} \ominus V(\mathcal{H}).$$

Lemma 7.2 *The sequence $\{S^n N_{S^*}\}$ consists of mutually orthogonal subspaces of $\mathcal{H}$.*

The proof of this lemma is clear: for any $k_1, k_2 \in N_{S^*}$ and $m \in \mathbb{N}$, we observe that

$$< S^m k_1, k_2 > = < k_1, (S^*)^m k_2 > = 0.$$

Theorem 7.3 (Wold's Theorem) *Let S be an isometric operator in a Hilbert space $\mathcal{H}$. Define*

$$\mathcal{H}_\infty = \bigcap_n S^n \mathcal{H},$$

and

$$\mathcal{H}_{shift} = N_{S^*} \oplus S N_{S^*} \oplus \cdots \oplus S^k N_{S^*} \oplus \cdots .$$

Then the following statements hold.

(1) The space $\mathcal{H}$ is decomposed into the orthogonal direct sum

$$\mathcal{H} = \mathcal{H}_\infty \oplus \mathcal{H}_{shift}.$$

(2)

$$\mathcal{H}_\infty = \{x \in \mathcal{H} : \|(S^*)^n x\| = \|x\|, \; \forall n \in \mathbb{N}\}.$$

(3) *The operator S restricted on $\mathcal{H}_\infty$ is a unitary operator, and S is a unilateral shift in the space $\mathcal{H}_{shift}$.*

Proof We sketch a proof of this theorem for the reader's convenience.

Let a vector $y \in \mathcal{H}$ be orthogonal to every subspace $S^k N_{S^*}$, $k = 0, 1, \ldots$. In particular, it follows from (7.1) that

$$y \in (N_{S^*})^\perp \iff y \in R_S \iff E_1 y = y.$$

It turns out that a more general result can be proved.

Lemma 7.4 *For any $n \in \mathbb{N}$, one has*

$$y \in (S^n N_{S^*})^\perp \iff E_{n+1} y = y.$$

Proof To see that the statement of this lemma is true, we apply the following sequence of equivalences:

$$y \in \left(S^n N_{S^*}\right)^\perp \iff \langle y, \; S^n x \rangle = 0 \qquad \forall x \in N_{S^*}$$

$$\iff \langle (S^*)^n y, \; x \rangle = 0 \qquad \forall x \in N_{S^*}$$

$$\iff (S^*)^n y \in R_S \qquad \text{(see (7.1))}$$

$$\iff \exists x \in \mathcal{H} \text{ such that } (S^*)^n y = Sx$$

$$\iff E_{n+1} y = y.$$

The last equivalence follows from the relation

$$E_{n+1} y = S^{n+1}(S^*)^{n+1} y = S^{n+1}(S^* S)x = S^{n+1} x = y$$

that proves the lemma.   □

We continue the proof of the theorem. It follows from Lemma 7.1 that the strong limit

$$\lim_{n \to \infty} E_n = E_\infty$$

exists, and is the projection onto the subspace

$$\mathcal{H}_\infty = \bigcap_n S^n \mathcal{H} = \bigcap_n (S^n N_{S^*})^\perp.$$

Next, we prove that S and S^* restricted to $\mathcal{H}_\infty$ are unitary operators. As a corollary, we obtain a few formulas involving these operators. For this, we show that

$$x \in \mathcal{H}_\infty \iff ||(S^*)^n x|| = ||x||, \quad \forall n \in \mathbb{N}.$$

Observe first that $||x||^2 = ||E_n x||^2 + ||E_n^\perp x||^2$ where $E_n^\perp = I_\mathcal{H} - E_n$ and $x \in \mathcal{H}, n \in \mathbb{N}$. Since $||E_n x - E_\infty x|| \to 0$ for all $x \in \mathcal{H}$, we obtain that

$$x \in \mathcal{H}_\infty \iff E_n^\perp x \to 0 \ (n \to \infty) \iff E_n^\perp x \to 0.$$

Because $||(S^*)^n x||^2 = ||S^n (S^*)^n x||^2 = ||E_n x||^2$, we conclude that

$$||(S^*)^n x||^2 = ||x|| \iff E_n^\perp x = 0.$$

Furthermore,

$$x \in \mathcal{H}_\infty \iff E_n^\perp x = 0 \ \forall n \in \mathbb{N} \iff ||(S^*)^n x|| = ||x||.$$

In particular, this means that $SS^*|_{\mathcal{H}_\infty} = I_{\mathcal{H}_\infty}$.

We notice that the subspace $\mathcal{H}_\infty$ is invariant with respect to S and S^*:

$$S^*(\mathcal{H}_\infty) \subset \mathcal{H}_\infty \iff S(\mathcal{H}_\infty^\perp) \subset \mathcal{H}_\infty^\perp.$$

Indeed, any vector x from $\mathcal{H}_\infty^\perp$ has the form

$$x = k_0 + S k_1 + \cdots + S^i k_i + \cdots$$

where all k_i are from N_{S^*}, and

$$S(k_0 + S k_1 + S^2 k_2 + \cdots) = 0 + S k_0 + S^2 k_1 + \cdots .$$

Thus, the operator S on $\mathcal{H}_\infty^\perp = \mathcal{H}_{shift}$ is a unilateral shift,

$$(k_0, k_1, k_2, \ldots) \mapsto (0, k_0, k_1, k_2, \ldots).$$

$\square$

7.2 Automorphic Factors and Exact Endomorphisms

The goal of this section is to apply the Wold theorem to the study of isometries generated by endomorphisms of a measure space.

We recall first the definition of a factor map. Let $(X, \mathcal{B})$ and $(Y, \mathcal{A})$ be standard Borel spaces, and let $\sigma : X \to X$ and $\tau : Y \to Y$ be surjective maps. It is said that $F : (X, \mathcal{B}, \sigma) \to (Y, \mathcal{A}, \tau)$ is a *factor map* if F is measurable, and $F \circ \sigma = \tau \circ F$.

Then τ is called a *factor* of σ. If τ is a Borel *automorphism*, then the dynamical system $(Y, \mathcal{A}, \tau)$ is called an *automorphic factor*. These definitions can be obviously reformulated in the context of measurable dynamical systems when σ and τ are non-singular (or measure preserving) maps.

Suppose ζ is a measurable partition of $(X, \mathcal{B}, \mu)$. Then we can define the quotient space

$$(Y, \mathcal{B}_\zeta, \mu_\zeta) = (X/\zeta, \mathcal{B}/\zeta, \mu/\zeta)$$

(see Sect. 2.3). Let $\phi : X \to Y$ be the natural projection. If, additionally, the partition ζ is invariant with respect to σ, i.e., $\sigma^{-1}\zeta \preceq \zeta$, then σ defines an onto endomorphism $\tilde{\sigma}$ of Y such that ϕ is a factor map: $\phi\sigma = \tilde{\sigma}\phi$.

To define an isometry generated by a surjective endomorphism σ, we assume that σ is a finite measure-preserving endomorphism of a standard measure space $(X, \mathcal{B}, \mu)$, and $\mu \circ \sigma^{-1} = \mu$. The assumption about the invariance of μ is not crucial and is made for convenience. The definition can be easily modified to the case of non-singular endomorphisms.

Theorem 7.5 *Let $(X, \mathcal{B}, \mu, \sigma)$ be a measure preserving non-invertible dynamical system. Let $\mathcal{H} = L^2(\mu)$ and define*

$$S : f \mapsto f \circ \sigma : \mathcal{H} \to \mathcal{H}.$$

Then S is an isometry. The adjoint of S is

$$S^* g = \frac{(g d\mu) \circ \sigma^{-1}}{d\mu}, \quad g \in \mathcal{H}.$$

Proof The fact that S is isometry follows from σ-invariance of μ. The formula for S^* is deduced as follows:

$$\langle Sf, g \rangle = \int_X (f \circ \sigma) g \, d\mu$$

$$= \int_X f(g d\mu) \circ \sigma^{-1}$$

$$= \int_X f \frac{(g d\mu) \circ \sigma^{-1}}{d\mu} d\mu$$

$$= \langle f, S^* g \rangle.$$

As was mentioned in Chap. 5, the co-isometry S^* is, in fact, a transfer operator R corresponding to the endomorphism σ.   □

It follows from this lemma that we can apply the Wold theorem for S and construct an orthogonal decomposition of $\mathcal{H} = L^2(\mu)$. It says that $\mathcal{H}$ can be decomposed into the orthogonal sum $\mathcal{H}_\infty \oplus \mathcal{H}_\infty^\perp$ where S restricted on $\mathcal{H}_\infty$ is a unitary operator. It turns out that the subspace $\mathcal{H}_\infty$ can be explicitly described in terms of the endomorphism σ.

We recall that, to every endomorphism σ of a Borel space $(X, \mathcal{B})$, one can associate two partitions of X. Let ξ be the partition of X into the σ-orbits, i.e., $\xi = \{Orb_\sigma(x) : x \in X\}$, where $y \in Orb_\sigma(x)$ if and only if there exist $m, n \in \mathbb{N}$ such that $\sigma^m(x) = \sigma^n(y)$. Define also a partition η of X into equivalence classes such that $x \sim y$ if and only if $\sigma^n(x) = \sigma^n(y)$ for some $n \in \mathbb{N}$. Then

$$\eta(x) = \bigcup_n \sigma^{-n}(\sigma^n(x)).$$

If σ is an at most countable-to-one endomorphism, then the partitions ξ and η define hyperfinite countable Borel equivalence relations (see [DJK94] for detail). Clearly, η-equivalence classes refine ξ-equivalence classes.

Suppose that $\mu \in M_1(X)$ is a σ-invariant measure, so that σ is considered as a measure preserving endomorphism of $(X, \mathcal{B}, \mu)$. We denote by ξ' and η' the measurable hulls of the partitions ξ and η, respectively.

It is worth noting that there exists a one-to-one correspondence between measurable partitions and complete sigma-subalgebras $\mathcal{A}$ of $\mathcal{B}$. Let $\mathcal{A}(\xi')$ and $\mathcal{A}(\eta')$ be the subalgebras corresponding to ξ' and η'.

For $(X, \mathcal{B}, \mu, \sigma)$ as above, define

$$\mathcal{B}_\infty = \bigcap_{n=0}^{\infty} \sigma^{-n}(\mathcal{B}),$$

and let $\mathcal{A}_\sigma = \{A \in \mathcal{B} : \sigma^{-1}(A) = A\}$ be the subalgebra of σ-invariant subsets of X. We recall that σ is called *exact* if $\mathcal{B}_\infty$ is a trivial subalgebra, and σ is called *ergodic* if $\mathcal{A}_\sigma$ is trivial. Since $\mathcal{A}_\sigma \subset \mathcal{B}_\infty$, exactness implies ergodicity.

If ϵ denotes the partition of X into points, then we have the sequence of decreasing measurable partitions $\{\sigma^{-i}\epsilon\}_{i=0}^\infty$:

$$\epsilon \geq \sigma^{-1}\epsilon \geq \sigma^{-2}\epsilon \cdots .$$

The objects, we have defined above, satisfy the following properties.

Lemma 7.6 ([Roh61]) *In the above notation, we have:*

(1)

$$\xi' \leq \eta', \qquad \eta' = \bigwedge_n \sigma^{-n}\epsilon;$$

(2)

$$A(\xi') = A_\sigma, \qquad A(\eta') = B_\infty.$$

In particular, it follows from Lemma 7.6 that an endomorphism σ is ergodic if the partition ξ' is trivial, and σ is exact if the partition η' is trivial (understood in terms of mod 0 convention).

Since η' is a measurable partition, we can define the quotient measure space $(X/\eta', B/\eta', \mu_{\eta'})$. By Lemma 7.6, we see that $B/\eta' = B_\infty$ and

$$Y := X'_\eta = X_{\bigwedge_n \sigma^{-n}\epsilon}.$$

Corollary 7.7

(1) Let $\phi : X \to Y$ be the natural projection. Then there exists a measure preserving automorphism $\tilde{\sigma} : (Y, \mu_{\eta'}) \to (Y, \mu_{\eta'})$ such that $\tilde{\sigma}$ is an automorphic factor of σ, i.e.,

$$\tilde{\sigma} \circ \phi = \phi \circ \sigma.$$

(2) Let $S : f \to f \circ \sigma$ be the isometry on $\mathcal{H} = L^2(\mu)$. Then, in the Wold decomposition $\mathcal{H} = \mathcal{H}_\infty \oplus \mathcal{H}_\infty^\perp$ for S, we have

$$\mathcal{H}_\infty = L^2(Y, \mu_{\eta'}),$$

and the restriction of S to $\mathcal{H}_\infty$ corresponds to the unitary operator U defined by $\tilde{\sigma}$, $U(f) = f \circ \tilde{\sigma}$.

Remark 7.8 Let σ be an endomorphism of a standard measure space (X, B, μ) as above. Then it follows from the construction of B_∞ and from the definition of the partition η' that for every B_∞-measurable function f there exists a sequence of functions (F_n) such that every F_n is B-measurable and, for every $n \in \mathbb{N}$,

$$f = F_n \circ \sigma^n. \tag{7.2}$$

With some abuse of notation, this relation can be also written as

$$\mathcal{M}_n(X) = \mathcal{M}(X) \circ \sigma^n,$$

where $\mathcal{M}_n(X)$ denotes the space of $\sigma^{-n}(B)$-measurable functions.

Moreover, one can show that a B_∞-measurable function f admits a representation $f = F_n \circ \sigma^n$ for every $n \in \mathbb{N}$ if and only if f is a constant function on every class of the measurable equivalence relation η'.

We consider now an application of relation (7.2) to transfer operators R defined on (X, B, μ) by an onto endomorphism σ. Suppose that σ is not exact, i.e., the subalgebra $\mathcal{B}_\infty$ is not trivial. By Remark 7.8 and relation (7.2), we can see that, for any $f \in \mathcal{M}(\mathcal{B}_\infty)$,

$$R^n(f) = F_n \omega_n, \quad \omega_n := R^n(1).$$

This fact can be easily proved by induction. Furthermore, since μ is σ-invariant, one can show that

$$F_n = (S^*)^n f,$$

where $S : f \mapsto f \circ \sigma$ is the isometry considered above. We leave the details to the reader.

Remark 7.9 In this section we have considered the case of measure preserving endomorphism σ. But the proved results are still true (mutatis mutandis) in the case when μ is non-singular with respect to σ. The standard method of dealing with non-singular transformations is as follows.

Let θ_μ be the Radon-Nikodym derivative, i.e. $\theta_\mu = \dfrac{d\mu \circ \sigma^{-1}}{d\mu}$. Then

$$S : f \mapsto \sqrt{\theta_\mu}(f \circ \sigma), \quad f \in L^2(\mu)$$

is an isometry in $\mathcal{H} = L^2(\mu)$. Hence one can use the arguments developed above in this section for the study of the operator S. In particular, the adjoint of S can be determined by formula

$$S^* g = \frac{(\sqrt{\theta_\mu} g \, d\mu) \circ \sigma^{-1}}{d\mu}.$$

Chapter 8
Operators on the Universal Hilbert Space Generated by Transfer Operators

Abstract Starting with a fixed transfer operator (R, σ) on $(X, \mathcal{B})$, we show below that there is then a naturally induced universal Hilbert space $\mathcal{H}(X)$ with the property that (R, σ) yields naturally a corresponding isometry in $\mathcal{H}(X)$, i.e., an isometry with respect to the inner product from $\mathcal{H}(X)$. With this, we then obtain a rich spectral theory for the transfer operators, for example a setting which may be considered to be an infinite-dimensional Perron-Frobenius theory. Our main results are Theorems 8.12, 8.17, and 8.18.

Keywords Spectral theory · The universal Hilbert space

8.1 Definition of the Universal Hilbert Space $\mathcal{H}(X)$

For the reader's convenience, we recall the definition of the *universal Hilbert space* $\mathcal{H}(X)$ where $(X, \mathcal{B})$ is a standard Borel space as usual. We will use [Nel69] as a main source. Our analysis of transfer operators in $\mathcal{H}(X)$ is motivated by [AJL16, Jor01, Jor04].

Let $M(X)$ be the set of all Borel measures on X. We write (f, μ) for a pair consisting of a measure $\mu \in M(X)$ and a function $f \in L^2(\mu)$.

Definition 8.1 It is said that two pairs (f, μ) and (g, ν) are *equivalent* if there exists a measure $\lambda \in M(X)$ such that $\mu \ll \lambda$ and $\nu \ll \lambda$, and

$$f\sqrt{\frac{d\mu}{d\lambda}} = g\sqrt{\frac{d\nu}{d\lambda}}, \quad \lambda\text{-a.e.} \qquad (8.1)$$

The set of equivalence classes of pairs (f, μ) is denoted by $\mathcal{H}(X)$.

© Springer International Publishing AG, part of Springer Nature 2018

S. Bezuglyi, P. E. T. Jorgensen, *Transfer Operators, Endomorphisms, and Measurable Partitions*, Lecture Notes in Mathematics 2217, https://doi.org/10.1007/978-3-319-92417-5_8

It is not hard to show that, if relation (8.1) holds for some λ, then

$$f\sqrt{\frac{d\mu}{d\lambda'}} = g\sqrt{\frac{d\nu}{d\lambda'}}, \quad \lambda'\text{-a.e.},$$

for any measure λ' such that $\mu \ll \lambda'$ and $\nu \ll \lambda'$ [Nel69]. From this observation, one can conclude that (8.1) defines an *equivalence relation* on the set of pairs (f, μ). We will denote the equivalence class of a pair (f, μ) by $f\sqrt{d\mu}$.

Remark 8.2

(1) We mention an important case of equivalence of two pairs, (f, μ) and (f', μ'). Suppose that $\mu' \ll \mu$ and $d\mu' = \varphi d\mu$. Then we can take $\lambda = \mu$ in Definition 8.1, so that

$$(f, \mu) \sim (f', \mu') \iff f = f'\sqrt{\varphi}, \quad \mu\text{-a.e.},$$

and these pairs belong to the class $f\sqrt{d\mu}$.

(2) Suppose that μ' is a probability measure equivalent to a given measure μ on X. Let $\varphi > 0$ be the Radon-Nikodym derivative $\dfrac{d\mu'}{d\mu}$. It follows from (1) that the two pairs (f, μ) and (f', μ') are equivalent if $f = f'\sqrt{\varphi}$. Hence, one can always assume that any equivalence class $f\sqrt{d\mu}$ is defined by a probability measure.

It turns out that $\mathcal{H}(X)$ can be endowed with a *Hilbert space* structure. To see that $\mathcal{H}(X)$ is a vector space, we define the sum of elements from $\mathcal{H}(X)$ as follows:

$$f\sqrt{d\mu} + g\sqrt{d\nu} = \left(f\sqrt{\frac{d\mu}{d\lambda}} + g\sqrt{\frac{d\nu}{d\lambda}} \right)\sqrt{d\lambda},$$

where $\mu \ll \lambda$ and $\nu \ll \lambda$ for some measure λ. The definition of the multiplication by a scalar is obvious. Next, an inner product on $\mathcal{H}(X)$ is defined by

$$\langle f\sqrt{d\mu}, \, g\sqrt{d\nu}\rangle_{\mathcal{H}(X)} = \int_X fg\sqrt{\frac{d\mu}{d\lambda}}\sqrt{\frac{d\nu}{d\lambda}}\, d\lambda \qquad (8.2)$$

where again $\mu \ll \lambda$ and $\nu \ll \lambda$ for a measure λ. It is a simple exercise to show that these operations are well-defined and do not depend on the choice of λ.

Proposition 8.3 ([Nel69]) *With respect to the operations defined above, $\mathcal{H}(X)$ is a Hilbert space.*

A proof of this assertion can be found in [Nel69] or [Jor04].

We will call $\mathcal{H}(X)$ the *universal Hilbert space*.

It follows from the definition of the inner product in $\mathcal{H}(X)$ that for any element $f\sqrt{d\mu}$ of $\mathcal{H}(X)$

$$\|f\sqrt{d\mu}\|^2_{\mathcal{H}(X)} = \int_X f^2\,d\mu = \|f\|^2_{L^2(\mu)}.$$

Thus, if μ is a fixed measure on X, then the map

$$\iota : f \mapsto f\sqrt{d\mu} : L^2(\mu) \to \mathcal{H}(X) \tag{8.3}$$

gives an *isometric embedding* of $L^2(\mu)$ into $\mathcal{H}(X)$.

We denote $\mathcal{H}(\mu) := \iota(L^2(\mu))$. The following proposition explains why $\mathcal{H}(X)$ is called a universal Hilbert space.

Proposition 8.4 *For any two measures μ and ν on $(X, \mathcal{B})$,*

(1) $\mu \ll \nu$ *if and only if $\mathcal{H}(\mu)$ is isometrically embedded into $\mathcal{H}(\nu)$.*
(2) $\mu \sim \nu$ *if and only if $\mathcal{H}(\mu) = \mathcal{H}(\nu)$.*
(3) μ *and* ν *are mutually singular if and only if the subspaces $\mathcal{H}(\mu)$ and $\mathcal{H}(\nu)$ are orthogonal in $\mathcal{H}(X)$.*

Proof These properties are rather obvious, and can be proved directly. We sketch here a proof of (1) to illustrate the used technique. Let $\psi d\nu = d\mu$. Set

$$T(g\sqrt{d\mu}) = g\sqrt{\psi}\sqrt{d\nu},$$

and show that T implements the isometric embedding. We have

$$\|T(g\sqrt{d\mu})\|^2_{\mathcal{H}(\nu)} = \int_X g^2\psi\,d\nu = \int_X g^2\,d\mu = \|g\sqrt{d\mu}\|^2_{\mathcal{H}(\mu)}.$$

We leave the proof of the other statements to the reader, see details in [Nel69, Jor04].
□

8.2 Transfer Operators on $\mathcal{H}(X)$

Suppose that we have a surjective endomorphism σ of a standard Borel space $(X, \mathcal{B})$, and let a transfer operator (R, σ) be defined on Borel functions on $(X, \mathcal{B})$. In our further considerations, we will work with the transfer operator R acting in the space $L^2(\lambda)$ where λ is a measure from $M(X)$. For given R, we divide measures into two subsets, $\mathcal{L}(R)$ and $M(X) \setminus \mathcal{L}(R)$.

We recall that a measure λ is called *atomic* if there exists a point in X of positive measure. Non-atomic measures are called *continuous*. Let $M_c(X)$ and $M_a(X)$ denote the subsets of $M(X)$ formed by continuous and purely atomic measures, respectively. Dealing with vectors $f\sqrt{d\lambda}$ in the space $\mathcal{H}(X)$, we distinguish two principal cases: (1) the measure λ is continuous or (2) the measure λ is purely atomic.

We discuss in the following statements some properties of the universal Hilbert space. Note that every measure μ can be viewed as a vector $\sqrt{d\mu}$ in the space $\mathcal{H}(X)$.

Lemma 8.5 *Let $\mathcal{G}$ be a subset of $M(X)$, and let $\mathcal{G}^{\perp}$ be the set of measures ν such that ν is singular to all measures μ from $\mathcal{G}$. Denote by $\mathcal{H}_{\mathcal{G}}$ the closure of the subspace spanned by $\mathcal{H}(\lambda)$, $\lambda \in \mathcal{G}$. The set $\mathcal{H}_{\mathcal{G}^{\perp}}$ is defined similarly. Then the universal Hilbert space $\mathcal{H}(X)$ admits the orthogonal decomposition:*

$$\mathcal{H}(X) = \mathcal{H}_{\mathcal{G}} \oplus \mathcal{H}_{\mathcal{G}^{\perp}}.$$

In other words, $(\mathcal{H}_{\mathcal{G}})^{\perp} = \mathcal{H}_{\mathcal{G}^{\perp}}$.

Lemma 8.5 follows immediately from Proposition 8.4.

Remark 8.6

(1) As mentioned in Remark 8.2, we can always assume that μ is a probability measure. Together with the assumption that $R(1) = 1$, this means that, for any probability measure μ, the measure μR is well defined. This fact will be repeatedly used below.
(2) For a specific choice of the set $\mathcal{G}$ in Lemma 8.5, we can get the following useful decompositions of $\mathcal{H}(X)$:

$$\mathcal{H}(X) = \mathcal{H}_{M_c(X)} \oplus \mathcal{H}_{M_a(X)}, \quad \mathcal{H}(X) = \mathcal{H}_{K_1} \oplus (\mathcal{H}_{K_1})^{\perp}, \tag{8.4}$$

$$\mathcal{H}(X) = \mathcal{H}_{\mathcal{L}(R)} \oplus (\mathcal{H}_{\mathcal{L}(R)})^{\perp} \tag{8.5}$$

where $K_1 = M_1(X)R$.
(3) Given a *nonzero* vector $f\sqrt{d\mu}$ in $\mathcal{H}(X)$ with a continuous (atomic) measure μ, we remark that the class of equivalent pairs generated by (f, μ) contains only pairs (g, λ) where λ is a continuous (atomic) measure. This follows from the following obvious fact: if $\lambda \ll \nu$ and λ is atomic at a point x_0, then ν is atomic at the same point. This means that $f\sqrt{d\mu} \in \mathcal{H}_{M_c(X)}$ if and only if μ is continuous, and $f\sqrt{d\mu} \in \mathcal{H}_{M_a(X)}$ if and only if μ is purely atomic. This means that the decomposition (8.4) is invariant with respect to the equivalence of pairs (f, μ). But the decomposition in (8.5) is not invariant with respect to this equivalence relation.

In what follows we will translate the notion of a transfer operator R and its adjoint operator S, which are studied in $L^2(\lambda)$ in Chap. 5, to the subspace $\mathcal{H}(\lambda)$ of the universal Hilbert space $\mathcal{H}(X)$. But, in contrast to the pair (R, S), we begin with an operator $\widehat{S}$ and show that its adjoint $\widehat{S}^* = \widehat{R}$ is an analogue of a transfer operator. Our approach is mainly based on the application of Proposition 8.4 which allows us to work with classes of equivalent measures.

In this section, we will deal with a pair of operators $(\widehat{R}, \widehat{S})$ acting in $\mathcal{H}(X)$ that are considered as analogous one to the symmetric pair of operators (R, S) studied in Chap. 5 where R is a transfer operator obtained as adjoint to the composition operator S. We first outline our approach to their definition. We recall that our main assumption in this context is that all considered transfer operators are normalized, $R(\mathbf{1}) = \mathbf{1}$.

We define an operator $\widehat{S}$ that acts in the set $\mathcal{P}$ of all pairs (f, μ) where $f \in L^2(\mu)$ and $\mu \in M_1(X)$. It will be checked that $\widehat{S}$ preserves the partition of $\mathcal{P}$ into equivalence classes. Therefore this fact allows us to consider $\widehat{S}$ as an operator acting in $\mathcal{H}(X)$. In the next step, we will check that $\widehat{S}$ is an isometry that leaves every subspace $\mathcal{H}(\lambda)$ invariant. Hence, the adjoint operator $\widehat{R} = \widehat{S}^*$ exists and is a co-isometry. We note that it is unclear whether $\widehat{R}$ can be defined directly in terms of a transformation on the set $\mathcal{P}$ that preserves the equivalence relation on the set of pairs (f, λ). Meantime, there exists a particular case when it can be done explicitly and this case will be studied carefully.

Given a vector $f \sqrt{d\lambda}$ with continuous measure $\lambda \in M_c(X)$, we are going to work with measures λR and $\lambda \circ \sigma^{-1}$. We can do it by virtue of Remark 8.6 (1). In other words, when we deal with pairs (f, λ), we can think that the actions of R and σ on the set of measures $M_1(X)$ are well defined everywhere. As was explained in Remark 8.6 (3), we can consider the two cases of continuous and purely atomic measures independently due to the invariance of the decomposition (8.4).

As discussed in Chap. 3, the assumption that $R(\mathbf{1}) = \mathbf{1}$ is not restrictive if a non-trivial harmonic function exists. On the other hand, this property is automatically true for a wide class of transfer operators acting in $L^2(\lambda)$ for any measure λ.

We begin with the following lemma which is used repeatedly below.

Lemma 8.7 *Let (R, σ) be a transfer operator on Borel functions over $(X, \mathcal{B})$. Then*

$$R(\mathbf{1}) = \mathbf{1} \quad \Longleftrightarrow \quad (\mu \circ R)\sigma^{-1} = \mu \ \forall \mu \in M_1(X).$$

This result immediately follows from the relation

$$R(\mathbf{1}) = \frac{d(\mu \circ R)\sigma^{-1}}{d\mu}$$

that was proved in Chap. 5, see (5.2).

Definition 8.8 Let λ be a continuous probability measure on $(X, \mathcal{B})$, $R(\mathbf{1}) = \mathbf{1}$, and $f \in L^2(\lambda)$. Then we define, for any pair (f, λ),

$$\widehat{S}(f, \lambda) = (f \circ \sigma, \lambda R). \tag{8.6}$$

We first show that the operator $\widehat{S}$ induces an operator on the space $\mathcal{H}(X)$. This fact follows from the following lemma.

Lemma 8.9 *Let $f \in L^2(\lambda)$ and $f_1 \in L^2(\lambda_1)$ where λ and λ_1 are continuous probability measures. Then*

$$(f_1, \lambda_1) \sim (f, \lambda) \iff (f_1 \circ \sigma, \lambda_1 R) \sim (f \circ \sigma, \lambda R)$$

Proof By definition of the equivalence relation $\sim$ on the set $\mathcal{P}$, two pairs (f, λ) and (f_1, λ_1) are in the same class if and only if there exists a measure μ such that $\lambda \ll \mu$, $\lambda_1 \ll \mu$, and

$$f \sqrt{\frac{d\lambda}{d\mu}} = f_1 \sqrt{\frac{d\lambda_1}{d\mu}}, \quad \mu\text{-a.e.} \tag{8.7}$$

In particular, μ can be chosen as the sum $\lambda + \lambda_1$. Then $\lambda R \ll \mu R$ and $\lambda_1 R \ll \mu R$ (see Chap. 5). It follows from Theorem 5.9 (1) that

$$\frac{d\lambda}{d\mu} \circ \sigma = \frac{d(\lambda R)}{d(\mu R)}, \qquad \frac{d\lambda_1}{d\mu} \circ \sigma = \frac{d(\lambda_1 R)}{d(\mu R)}. \tag{8.8}$$

Hence, we can apply (8.7), (8.8) and conclude that

$$(f \circ \sigma)\sqrt{\frac{d(\lambda R)}{d(\mu R)}} = (f \circ \sigma)\sqrt{\frac{d\lambda}{d\mu}} \circ \sigma$$

$$= (f_1 \circ \sigma)\sqrt{\frac{d\lambda_1}{d\mu}} \circ \sigma$$

$$= (f_1 \circ \sigma)\sqrt{\frac{d(\lambda_1 R)}{d(\mu R)}} \qquad (\mu R)\text{-a.e.}$$

This proves that $(f_1 \circ \sigma, \lambda_1 R) \sim (f \circ \sigma, \lambda R)$.

Conversely, if we have the fact that the pairs $(f_1 \circ \sigma, \lambda_1 R)$ and $(f \circ \sigma, \lambda R)$ are equivalent, then

$$(f \circ \sigma)\sqrt{\frac{d(\lambda R)}{d(\mu R)}} = (f_1 \circ \sigma)\sqrt{\frac{d(\lambda_1 R)}{d(\mu R)}}, \qquad (\mu R)\text{-a.e.}$$

Hence, we can apply the transfer operator R to the both sides of this relation, and because $\mu R \ll \mu$, we obtain (8.7). $\qquad \square$

Remark 8.10 If a measure λ is in the set $\mathcal{L}(R)$, then, for some measurable function W, we have $d(\lambda R) = W d\lambda$, Then the operator $\widehat{S}$ acts in the subspace $\mathcal{H}(\lambda)$ as follows:

$$\widehat{S}(f\sqrt{d\lambda}) = (f \circ \sigma)\sqrt{W}\sqrt{d\lambda}. \tag{8.9}$$

In order to justify (8.9), we observe that if $(f, \lambda) \sim (f, \lambda_1)$ with $\lambda, \lambda_1 \in \mathcal{L}(R)$, then

$$(\sqrt{W}(f \circ \sigma), \lambda) \sim (\sqrt{W_1}(f_1 \circ \sigma), \lambda_1)$$

where $W_1 d\lambda_1 = d(\lambda_1 R)$. This equivalence can be directly deduced from the relation $W_1 = (\varphi \circ \sigma)W\varphi^{-1}$ where $\varphi d\lambda = d\lambda_1$ that was discussed in Chap. 5.

In the next statement, we show that the operator $\widehat{S}$, defined in (8.6), factors to an operator in the universal Hilbert space.

Lemma 8.11 *The operator*

$$\widehat{S}(f\sqrt{d\lambda}) = (f \circ \sigma)\sqrt{d(\lambda R)}, \tag{8.10}$$

is well defined in the universal Hilbert space $\mathcal{H}(X)$. *Furthermore,* $\widehat{S}$ *is bounded if and only if* $R(\mathbf{1}) \in L^\infty(\lambda)$.

We use the same notation $\widehat{S}$ for the operators acting on the set of pairs (f, λ) and in the Hilbert space $\mathcal{H}(X)$. It will be clear from the context where $\widehat{S}$ acts.

Proof We first need to justify the correctness of the definition $\widehat{S}$. Indeed, this result follows from Lemma 8.9 because if we take any two pairs (f, λ), (f_1, λ_1) that belong to the same class, then $\widehat{S}$ maps these pairs into equivalent pairs $(f \circ \sigma, \lambda R)$ and $(f_1 \circ \sigma, \lambda_1 R)$. Hence relation (8.10) defines a transformation in $\mathcal{H}(X)$.

To see that $\widehat{S}$ is a linear operator we have to check that $\widehat{S}(c_1 f\sqrt{d\lambda} + c_2 f_1\sqrt{d\lambda_1}) = c_1\widehat{S}(f\sqrt{d\lambda}) + c_2\widehat{S}(f_1\sqrt{d\lambda_1})$. This can be proved again by the choice of representatives in the classes $f\sqrt{d\lambda}$ and $f_1\sqrt{d\lambda_1}$ as we did above. The details are left to the reader. $\square$

We recall that $\mathcal{H}(\lambda)$ denotes the subspace of $\mathcal{H}(X)$ obtained by the isometric embedding of $L^2(\lambda)$ into $\mathcal{H}(X)$.

Theorem 8.12 *Let* (R, σ) *be a transfer operator on* $(X, \mathcal{B})$. *Then the operator* $\widehat{S}$ *of* $\mathcal{H}(X)$ *is an isometry if and only if* $R(\mathbf{1}) = \mathbf{1}$. *Moreover, if a measure* $\lambda \in \mathcal{L}(R)$, *then the subspace* $\mathcal{H}(\lambda)$ *is invariant with respect to* $\widehat{S}$.

Proof To see that $\widehat{S}$ is an isometry, we use (8.2) and calculate

$$\|\widehat{S}(f\sqrt{d\lambda})\|_{\mathcal{H}(X)}^2 = \int_X (f \circ \sigma)^2 \, d(\lambda R)$$

$$= \int_X R[(f \circ \sigma)^2] \, d\lambda$$

$$= \int_X f^2 R(\mathbf{1}) \, d\lambda.$$

Hence, we see that

$$\|\widehat{S}(f\sqrt{d\lambda})\|_{\mathcal{H}(X)}^2 = \|f\|_{\mathcal{H}(X)}^2$$

if and only if $R(\mathbf{1}) = \mathbf{1}$.

To prove the second part of the theorem, we suppose that $\lambda \in \mathcal{L}(R)$, then $d(\lambda R) = W d\lambda$. Take any element $f\sqrt{d\mu} \in \mathcal{H}(\lambda)$. By Proposition 8.4, this means that $\mu \sim \lambda$, $d\mu = \varphi d\lambda$, and (f, μ) is equivalent to (g, λ). Then $d(\mu R) = (\varphi \circ \sigma) W \varphi^{-1} d\mu$ (see Chap. 5). It follows from this fact that

$$\widehat{S}(f\sqrt{d\mu}) = (f \circ \sigma)\sqrt{d(\mu R)} = [\sqrt{W}(f\varphi) \circ \sigma]\sqrt{d\lambda}.$$

It follows from the definition of equivalence of pairs (f, μ) and (g, λ) that

$$\sqrt{W}(f\varphi) \circ \sigma] \in L^2(\lambda).$$

Hence $\widehat{S} : \mathcal{H}(\lambda) \to \mathcal{H}(\lambda)$, and the theorem is proved.   □

We can immediately deduce from Theorem 8.12 several important properties of $\widehat{S}$ and its adjoint. We recall that the transfer operator R defines an action on the set of probability measures, see Chap. 6. We use the notation $K_1 := M_1 R$ for the set of measures of the form λR, $\lambda \in M_1$. Then the subspace $\mathcal{H}_{K_1}$ is spanned by $\{\mathcal{H}(\lambda) : \lambda \in K_1\}$.

Corollary 8.13

(1) *The decomposition $\mathcal{H}(X) = \mathcal{H}_{K_1} \oplus (\mathcal{H}_{K_1})^\perp$ implies that $\widehat{S}(\mathcal{H}(X)) = \mathcal{H}_{K_1}$.*
(2) *The adjoint operator $\widehat{S}^* : \mathcal{H}(X) \to \mathcal{H}(X)$ is well defined and $Ker(\widehat{S}^*) = (\mathcal{H}_{K_1})^\perp$.*

We remark that the adjoint operator $\widehat{S}^*$ is defined in terms of the Hilbert space $\mathcal{H}(X)$, in contrast to the case of $\widehat{S}$ where we first defined $\widehat{S}$ on the set of pairs (f, λ) and then extended to the classes of equivalence that form the Hilbert space $\mathcal{H}(X)$.

The next result gives an explicit formula for the action of $\widehat{S}^*$ when $\lambda R \ll \lambda$. We recall that with this assumption $\widehat{S}^*$ leaves the subspace $\mathcal{H}(\lambda)$ invariant.

Proposition 8.14 *Let (R, σ) be a normalized transfer operator and $\lambda \in \mathcal{L}(R)$. Then the adjoint operator $\widehat{S}^*$ acts on $\mathcal{H}(\lambda)$ by the formula:*

$$\widehat{S}^*(f\sqrt{d\lambda}) = R\left(\frac{f}{\sqrt{W}}\right)\sqrt{d\lambda} \tag{8.11}$$

where $W d\lambda = d(\lambda R)$.

Proof The result is proved by the following calculation:

$$\langle \widehat{S}(f\sqrt{d\lambda}), g\sqrt{d\lambda}\rangle_{\mathcal{H}(\lambda)} = \int_X \sqrt{W}(f \circ \sigma)g \, d\lambda \qquad \text{(see Remark 8.2)}$$

$$= \int_X (f \circ \sigma)g \frac{1}{\sqrt{W}} \, d(\lambda R)$$

$$= \int_X R\left((f \circ \sigma)g \frac{1}{\sqrt{W}}\right) d\lambda$$

$$= \int_X fR\left(\frac{g}{\sqrt{W}}\right) d\lambda$$

$$= \langle f\sqrt{d\lambda}, R(gW^{-1/2})\rangle_{\mathcal{H}(\lambda)}$$

and the proof is complete.   □

Corollary 8.15 *Let $\lambda \in \mathcal{L}(R)$. In the notation of Proposition 8.14, $\widetilde{\widehat{S}\widehat{S}}^*$ is the projection in the space $\mathcal{H}(\lambda)$ which acts by the formula:*

$$\widehat{S}\widehat{S}^*(f\sqrt{d\lambda}) = [R(\frac{f}{\sqrt{W}}) \circ \sigma]\sqrt{W}\sqrt{d\lambda}.$$

Proof This relation is proved by direct application of (8.10) and (8.11).   □

We return to the question about an explicit definition of the adjoint operator $\widehat{S}^*$. The key point is that the range of the isometry $\widehat{S}$ is the subspace $\mathcal{H}_{K_1}$, so that the kernel of $\widehat{S}^*$ must be $(\mathcal{H}_{K_1})^\perp$. In other words, $\widehat{S}^*(f\sqrt{d\lambda}) = 0$ if $\sqrt{\lambda} \in (\mathcal{H}_{K_1})^\perp$ according to Corollary 8.13. Here $\sqrt{\lambda}$ is considered as a vector in $\mathcal{H}(X)$.

To describe the action of $\widehat{S}^*$, we define an operator $\widehat{R}$ in the Hilbert space $\mathcal{H}(X)$ that is generated by the transfer operator R.

Definition 8.16 Let (R, σ) be a normalized transfer operator. We set, for any $f\sqrt{d\lambda}$,

$$\widehat{R}(f\sqrt{d\lambda}) = \begin{cases} R(f)\sqrt{d(\lambda \circ \sigma^{-1})}, & \sqrt{\lambda} \in \mathcal{H}_{K_1} \\ 0, & \sqrt{\lambda} \in (\mathcal{H}_{K_1})^\perp. \end{cases}$$

Because $\mathcal{H}(X) = \mathcal{H}_{K_1} \oplus (\mathcal{H}_{K_1})^{\perp}$, the operator $\widehat{R}$ is well-defined in $\mathcal{H}(X)$. With some abuse of notation, we will equally use the relation $\lambda \in K_1$ in the same meaning as $\sqrt{\lambda} \in \mathcal{H}_{K_1}$.

We remark that if $\lambda \circ \sigma^{-1} = \lambda$, then the operator $\widehat{R}$ sends $f\sqrt{d\lambda}$ to $R(f)\sqrt{d\lambda}$, and it can be identified with the transfer operator in R acting in $L^2(\lambda)$.

The following theorem is complimentary to the results obtained in Chap. 6. This theorem clarifies the role of the subset $K_1 \subset M(X)$.

Theorem 8.17 *For a normalized transfer operator (R, σ), the following statements are equivalent:*

(1) $\lambda \in K_1$;
(2) $(\lambda \circ \sigma^{-1})R = \lambda$;
(3) the map $f \mapsto R(f) \circ \sigma|_{L^2(\lambda)} = \mathbb{E}_{\lambda}(f \,|\sigma^{-1}(\mathcal{B}))$ where $\mathbb{E}_{\lambda}(\cdot \,|\sigma^{-1}(\mathcal{B}))$ is the conditional expectation on the subalgebra of $\sigma^{-1}(\mathcal{B})$-measurable functions in $L^2(\lambda)$;
(4) the operator $\widehat{E}_1 = \widehat{S}\widehat{R}$ maps $\mathcal{H}(\lambda)$ into itself, and

$$\widehat{E}_1(f\sqrt{d\lambda}) = \mathbb{E}_{\lambda}(f \,| \sigma^{-1}(\mathcal{B}))\sqrt{d\lambda}.$$

Proof The equivalence of statements (1) and (2) was proved in Theorem 6.9. Moreover, these two assertions are equivalent to the fact that the equation $\nu R = \lambda$ has a unique solution for every fixed $\lambda \in K_1$.

Suppose now that (1) and/or (2) hold. To show that (3) is true, we observe that the operator $P_{\lambda} = f \mapsto R(f) \circ \sigma|_{L^2(\lambda)}$ is obviously a projection in $L^2(\lambda)$ since $P_{\lambda}^2 = P_{\lambda}$ and $P_{\lambda}(g \circ \sigma) = g \circ \sigma$. It remains to show that $P_{\lambda} = P_{\lambda}^*$ or

$$\langle P_{\lambda} f_1, f_2 \rangle_{L^2(\lambda)} = \langle f_1, P_{\lambda} f_2 \rangle_{L^2(\lambda)}.$$

To see this, we compute, using that $\lambda = \nu R$,

$$\langle P_{\lambda} f_1, f_2 \rangle_{L^2(\lambda)} = \int_X (R(f_1) \circ \sigma) f_2 \, d\lambda$$

$$= \int_X (R(f_1) \circ \sigma) f_2 \, d(\nu R) \qquad (8.12)$$

$$= \int_X R[(R(f_1) \circ \sigma) f_2] \, d\nu$$

$$= \int_X R(f_1) R(f_2) \, d\nu.$$

By symmetry, we see that relation (8.12) gives also

$$\langle f_1, P_{\lambda} f_2 \rangle_{L^2(\lambda)} = \int_X R(f_1) R(f_2) \, d\nu.$$

Thus, P_λ is self-adjoint. We conclude that P_λ is the conditional expectation $\mathbb{E}_\lambda(\cdot \,|\sigma^{-1}(\mathcal{B}))$.

(3) $\implies$ (4) We apply $\widehat{E}_1$ to a vector $(f\sqrt{d\lambda}) \in \mathcal{H}(\lambda)$ and find

$$(\widehat{S}\widehat{R})(f\sqrt{d\lambda}) = \widehat{S}(R(f)\sqrt{d(\lambda \circ \sigma^{-1})})$$

$$= (R(f)\circ\sigma)\sqrt{d(\lambda \circ \sigma^{-1})R}$$

$$= (R(f)\circ\sigma)\sqrt{d\lambda}.$$

The result then follows from (3).

(4) $\implies$ (1) The operator $\widehat{R}$ is nonzero on the elements of $f\sqrt{d\lambda} \in \mathcal{H}(X)$ if and only if λ is in K_1.

$\square$

Theorem 8.18 *Let (R, σ) be a transfer operator such that $R(1) = 1$. The operators $\widehat{R}$ and $\widehat{S}$ form a symmetric pair in $\mathcal{H}(X)$, that is $\widehat{R} = \widehat{S}^*$.*

Proof We need to show that

$$\langle \widehat{S}(f\sqrt{d\nu}), g\sqrt{d\mu}\rangle_{\mathcal{H}(X)} = \langle f\sqrt{d\nu}, R(g)\sqrt{d(\mu \circ \sigma^{-1})}\rangle_{\mathcal{H}(X)} \qquad (8.13)$$

It suffices to assume that $\sqrt{d\mu} \in \mathcal{H}_{K_1}$ because for $\sqrt{d\mu} \in (\mathcal{H}_{K_1})^\perp$ the both parts of (8.13) are zeros.

Then, by Theorem 8.17, there exists a measure λ such that $d(\nu R) \ll d(\lambda R)$ and $d\mu = d(\mu \circ \sigma^{-1})R \ll d(\lambda R)$. We will use in the following computation the formulas that were proved in Chap. 5

$$\frac{d(\nu R)}{d(\lambda R)} = \frac{d\nu}{d\lambda} \circ \sigma,$$

$$\frac{d(\mu)}{d(\lambda R)} = \frac{d((\mu \circ \sigma^{-1})R)}{d(\lambda R)} = \frac{d\mu \circ \sigma^{-1}}{d\lambda} \circ \sigma.$$

Thus, we have

$$\langle \widehat{S}(f\sqrt{d\nu}), g\sqrt{d\mu}\rangle_{\mathcal{H}(X)} = \int_X (f\circ\sigma)g\sqrt{\frac{d(\nu R)}{d(\lambda R)}}\sqrt{\frac{d(\mu)}{d(\lambda R)}}\, d(\lambda R)$$

$$= \int_X (f\circ\sigma)g\sqrt{\frac{d\nu}{d\lambda}\circ\sigma}\sqrt{\frac{d(\mu \circ \sigma^{-1})}{d\lambda R}}\circ\sigma\, d(\lambda R)$$

$$= \int_X fR(g)\sqrt{\frac{d\nu}{d\lambda}}\sqrt{\frac{d(\mu \circ \sigma^{-1})}{d\lambda R}}\, d\lambda$$

$$= \langle f\sqrt{d\nu}, R(g)\sqrt{d(\mu \circ \sigma^{-1})}\rangle_{\mathcal{H}(X)}.$$

The proof is complete.

$\square$

Remark 8.19 In this remark, we collect a few facts about the operators $\widehat{S}$ and $\widehat{R}$.

(1) If the transfer operator (R, σ) is normalized, then It can be deduced directly from the definitions of the operators $\widehat{S}$ and $\widehat{R}$ that $\widehat{RS} = I_{\mathcal{H}(X)}$. Indeed, we have

$$(\widehat{RS})(f\sqrt{d\lambda}) = \widehat{R}((f \circ \sigma)\sqrt{d(\lambda R)}$$

$$= R(f \circ \sigma)\sqrt{d(\lambda R \circ \sigma^{-1})}$$

$$= f\sqrt{d\lambda}$$

where statement (2) of Theorem 8.17 was used.

(2) If R is not normalized, then $\widehat{S}$ is bounded in $\mathcal{H}(X)$ if and only if $R(1) \in L^\infty(\lambda)$ for all λ.

   To see this, we compute

$$\|\widehat{S}(f\sqrt{d\lambda})\|_{\mathcal{H}(X)}^2 = \int_X (f \circ \sigma)^2 \, d(\lambda \circ R) = \int_X f^2 R(1) \, d\lambda.$$

(3) The following result which is similar to Theorem 8.12 can be proved:
   Suppose h is a harmonic function for the transfer operator R acting in $L^2(\lambda)$. Then $\widehat{S}$ is an isometry in $L^2(hd\lambda)$.

The next lemma deals with non-normalized transfer operators (R, σ).

Lemma 8.20 *Let (R, σ) be a transfer operator. Suppose the operators $\widehat{R}$ and $\widehat{S}$ are defined as above. Then the operator $\widehat{RS}$ is a multiplication operator in $\mathcal{H}(X)$.*

Proof We obtain that

$$\widehat{RS}(f\sqrt{\lambda}) = \widehat{R}((f \circ \sigma)\sqrt{\lambda R}) = (R(1)f\sqrt{(\lambda R)} \circ \sigma^{-1}\sigma^{-1}) = (R(1)^{3/2}f\sqrt{\lambda}).$$

We used here relation (5.2).

□

Chapter 9
Transfer Operators with a Riesz Property

Abstract A well known theorem (Riesz) in analysis states that every positive linear functional L on continuous functions is represented by a Borel measure. More precisely, let X be a locally compact Hausdorff space and $C_c(X)$ the space of continuous functions with compact support. Then the well-known Riesz representation theorem from analysis says that, for every positive linear functional L, there exists a unique regular Borel measure μ on X such that

$$L(f) = \int_X f \, d\mu.$$

We are interested in a special case of functionals L_x defined on a function space by the formula $L_x(f) = f(x)$. For Borel functions $\mathcal{F}(X, \mathcal{B})$ over a standard Borel space $(X, \mathcal{B})$, the Riesz theorem is not directly applicable. We introduce in this chapter a class of transfer operators R that have the following property.

Keywords Riesz theorem · Riesz property · Riesz family · Set of probability measures

We begin with the following definition.

Definition 9.1 Let R be a positive operator acting on Borel functions over a standard Borel space $(X, \mathcal{B})$. We say that R has the *Riesz property* if, for every $x \in X$, there exists a Borel measure μ_x such that

$$R(f)(x) = \int_X f(y) \, d\mu_x(y), \qquad f \in \mathcal{F}(X, \mathcal{B}). \tag{9.1}$$

We call (μ_x) a *Riesz family* of measures corresponding to the operator R with Riesz property.

© Springer International Publishing AG, part of Springer Nature 2018
S. Bezuglyi, P. E. T. Jorgensen, *Transfer Operators, Endomorphisms,
and Measurable Partitions*, Lecture Notes in Mathematics 2217,
https://doi.org/10.1007/978-3-319-92417-5_9

In the following remark we present several facts that immediately follow from this definition.

Remark 9.2

(1) It follows from the equality $\mu_x(X) = R(\mathbf{1})(x)$ that μ_x is a probability measure for all $x \in X$ if and only if R is normalized.
(2) The field of measures $x \mapsto \mu_x$ is Borel in the sense that, for any Borel function $f \in \mathcal{F}(X, \mathcal{B})$, the function $x \mapsto \mu_x(f)$ is Borel. Indeed, this observation follows from (9.1) because $\mu_x(f) = R(f)(x)$.
(3) Given a positive operator R, the corresponding Riesz family (μ_x) is uniquely determined.

Suppose R is a positive operator with the Riesz property. Then any power R^k also has the Riesz property. So that we can write down for $f \in \mathcal{F}(X)$

$$R^k(f) = \int_X f \, d\nu_x^k, \qquad k \in \mathbb{N}.$$

On the other hand, if we iterate relation (9.1), then we obtain the following formula

$$R^k(f)(x) = \int_X \cdots \int_X f(y_k) \, d\mu_{y_{k-1}}(y_k) \cdots d\mu_x(y_1).$$

By uniqueness of the Riesz family, we conclude that

$$d\nu_x^k = \int_X \cdots \int_X d\mu_{y_{k-1}}(y_k) \cdots d\mu_x(y_1).$$

We will also write $d\mu_x(y) = d\mu(y|x)$ and treat this measure as conditional one. This point of view will be used for the case when all measures (μ_x) are pairwise singular.

So far, we have used only the property of positivity of the operator R. From now on, we will assume that R has the pull-out property, i.e., R is a transfer operator on $\mathcal{F}(X, \mathcal{B})$ corresponding to an onto endomorphism σ.

Lemma 9.3 *Suppose that (R, σ) is a transfer operator defined on $\mathcal{F}(X, \mathcal{B})$ such that $R(\mathbf{1}) = \mathbf{1}$. Assume that R has the Riesz property. Then, for the Riesz family of measures (μ_x), we have*

$$\mu_x \circ \sigma^{-1} = \delta_x, \qquad x \in X,$$

where δ_x is the Dirac measure.

Proof Since $\delta_x(f) = f(x)$, we note that the relation $\mu_x \circ \sigma^{-1} = \delta_x$ is equivalent to

$$\int_X f \, d(\mu_x \circ \sigma^{-1}) = f(x), \qquad \forall f \in \mathcal{F}(X, \mathcal{B}),$$

or, in other words, is equivalent to

$$\int_X (f \circ \sigma) \, d\mu_x = f(x), \qquad \forall f \in \mathcal{F}(X, \mathcal{B}).$$

But by (9.1), we obtain

$$\int_X (f \circ \sigma) \, d\mu_x = R(f \circ \sigma)(x)$$
$$= f(x)R(1)(x)$$
$$= f(x),$$

and we are done.   □

The following observation follows directly from this result.

Corollary 9.4 *Let* (R, σ) *be a transfer operator acting on* $\mathcal{F}(X, \mathcal{B})$. *Suppose that* R *has the Riesz property and* $R(1) = 1$. *Then, for any* $x \in X$,

$$\mathrm{supp}(\{\mu_x\}) = \sigma^{-1}(x),$$

where (μ_x) *is the Riesz family of measures corresponding to* R.

Lemma 9.5 *Let* R *be a transfer operator with Riesz property such that* $R(1) = 1$. *Suppose that*

$$\int_X f \, d\mu_x = R(f)(x)$$

for all $x \in X$. *Take a Borel measure* λ *on* $(X, \mathcal{B})$. *If* $\mathcal{H}(\mu_x)$ *is a subspace of the universal Hilbert space* $\mathcal{H}(X)$, *then the following statements are equivalent:*

(1)

$$\lambda \ll \mu_x, \qquad x \in X;$$

(2)

$$\mathcal{H}(\lambda) \hookrightarrow \mathcal{H}(\mu_x), \qquad x \in X;$$

(3)

$$\int_X f \, d\lambda = R\left(f \frac{d\lambda}{d\mu_x}\right)(x).$$

Proof The equivalence of (1) and (2) is mentioned in Proposition 8.4. The equivalence of these statements to (3) follows from the definition of the Riesz property. □

Lemma 9.6 *Let R be a transfer operator with Riesz property such that $R(\mathbf{1}) = \mathbf{1}$. If (μ_x) is the corresponding family of measures for R, then, for any sets A, $B \in \mathcal{B}$,*

$$\mu_x(\sigma^{-1}(A) \cap B) = \delta_x(A)\mu_x(B), \qquad x \in X.$$

Proof To show this, we use Definition 9.1 and Lemma 9.3. We compute

$$\begin{aligned}
\mu_x(\sigma^{-1}(A) \cap B) &= \int_X \chi_{\sigma^{-1}(A)} \chi_B \, d\mu_x \\
&= R(\chi_A \circ \sigma \, \chi_B)(x) \\
&= \chi_A(x) R(\chi_B)(x) \\
&= \delta_x(A)\mu_x(B)
\end{aligned}$$

□

In a similar way, we can formulate a simple general criterion for a positive operator R, defined by (9.1), to have the pull-out property.

Lemma 9.7 *A positive operator R with Riesz property is a transfer operator with pull-out property if and only if, for any measurable functions f, g from the domain of R,*

$$\int_X (f \circ \sigma) g \, d\mu_x = f(x) \int_X g \, d\mu_x.$$

Let (R, σ) be a transfer operator on Borel function $\mathcal{F}(X, \mathcal{B})$. We recall the construction of the induced transfer operator R_h where h is a positive harmonic function for R. Then

$$R_h(f) := \frac{R(fh)}{h}, \qquad f \in \mathcal{F}(X, \mathcal{B}), \tag{9.2}$$

is a transfer operator such that $R_h(\mathbf{1}) = \mathbf{1}$, i.e., R_h is normalized.

Proposition 9.8 *Given a transfer operators (R, σ) and an R-harmonic function h, let (R_h, σ) be defined as above. Suppose that R has the Riesz property, and let (μ_x) be the corresponding Riesz family of measures. Then R_h also has the Riesz* *property with respect the family* (μ'_x) *where the measures* (μ_x) *and* (μ'_x) *are related as follows:*

$$d\mu_x(y) = \frac{h(\sigma y)}{h(y)} d\mu'_x(y), \qquad y \in \sigma^{-1}(x),$$

In other words, the statement of the Proposition 9.8 says that the function $\dfrac{d\mu_x}{d\mu'_x}$ is a σ-coboundary.

Proof We need to find the family of measures (μ'_x) such that

$$R_h(f)(x) = \int_X f(y)\, d\mu'_x(y).$$

Since $R_h(f)$ can be found from (9.2), we can write

$$R_h(f) = \frac{R(fh)}{h}(x)$$

$$= R\left(\frac{fh}{h \circ \sigma}\right)(x)$$

$$= \int_X \frac{fh}{h \circ \sigma}(y)\, d\mu_x(y)$$

$$= \int_X f\, d\mu'_x.$$

Hence, we can take

$$d\mu'_x(y) = \frac{h}{h \circ \sigma}(y)\, d\mu_x(y).$$

We note that $h(x)d\mu'_x(y) = h(y)d\mu_x(y)$ for any x and $y \in \sigma^{-1}(x)$. $\square$

Let now v be a probability measure on $(X, \mathcal{B})$. Define a new measure λ on $(X, \mathcal{B})$ by the formula

$$\lambda = \int_X \mu_x\, dv(x).$$

This is equivalent to the equality

$$\int_X f\, d\lambda = \int_X \left(\int_X f(y)\, d\mu_x(y)\right) dv(x) \qquad (9.3)$$

which is used in the following statement.

We note that, in the case when a transfer operator R satisfies the Riesz property, the family of Riesz measures (μ_x) can be viewed as a system of conditional measures defined $(X, \mathcal{B}, \lambda)$ by the measurable partition $\xi = \{\sigma^{-1}(x) | x \in X\}$.

Proposition 9.9 *Let a transfer operators (R, σ) have the Riesz property with the family of measures (μ_x).*

(1) Let λ be a measure defined by v and (μ_x) as in (9.3). Then $\lambda = vR$.
(2) Suppose that $R(1) = 1$. Then, for v on $(X, \mathcal{B})$ and λ as above, we have $\lambda R = \lambda$.

Proof

(1) It follows from the definition of λ that, for any function f,

$$\int_X f(x)\, d\lambda = \int_X R(f)(x)\, dv$$

$$= \int_X f(x)\, d(vR)$$

and we are done.

(2) We need to show that for any function f the following equality holds

$$\int_X f\, d\lambda = \int_X f\, d(\lambda R)$$

$$= \int_X R(f)\, d\lambda.$$

In the following computation we use relation (9.3) and the fact that μ_x is a probability measure for all $x \in X$. By Definition 9.1, we have

$$\int_X R(f)(x)\, d\lambda(x) = \int_X \left(\int_X R(f)(y)\, d\mu_x(y) \right) d\lambda(x) \qquad \text{(by (9.3))}$$

$$= \int_{(x)} \left[\int_{(z)} \left(\int_{(y)} f(y)\, d\mu_x(y) \right) d\mu_x(z) \right] dv(x) \qquad \text{(by (9.1))}$$

$$= \int_{(x)} \left[\int_{(y)} f(y) \left(\int_{(z)} d\mu_x(z) \right) d\mu_x(y) \right] dv(x)$$

$$= \int_{(x)} \left[\int_{(y)} f(y)\, d\mu_x(y) \right] dv(x)$$

$$= \int_X f(x) d\lambda(x).$$

$\square$

Corollary 9.10 *Let R, (μ_x), ν and λ be as in Proposition 9.9. Suppose $R(\mathbf{1})(x) = W(x)$. Then*

$$W(x) = \frac{d(\lambda R)}{d\lambda}(x), \qquad x \in X.$$

Proof This result follows from the proof of Proposition 9.9 in which we will need to use the relation

$$W(x) = \int_X d\mu_x.$$

$\square$

Chapter 10
Transfer Operators on the Space of Densities

Abstract This chapter is focused on the study of an important class of transfer operators. As usual, we fix a non-invertible non-singular dynamical system $(X, \mathcal{B}, \mu, \sigma)$. Without loss of generality, we can assume that μ is a finite (even probability) measure because μ can be replaced by any measure equivalent to μ.

Keywords Non-invertible non-singular dynamical systems · Quasi-invariant measures

We recall that if λ is a Borel measure such that $\lambda \ll \mu$, then there exists the Radon-Nikodym derivative $f(x) = \dfrac{d\lambda}{d\mu}(x)$. Conversely, any nonnegative function $f \in L^1(\mu)$ serves as a density function for a measure $d\lambda = f d\mu$.

Definition 10.1 Define a transfer operator $R_\mu = (R, \sigma)$ acting on $L^1(\mu)$ by the formula

$$R_\mu(f)(x) = \frac{(f d\mu) \circ \sigma^{-1}}{d\mu}(x), \quad f \in L^1(\mu). \tag{10.1}$$

We call R_μ a transfer operator on the space of densities.

In this chapter, we will work only with transfer operators R_μ defined by (10.1).

The following lemma contains main properties of $R = R_\mu$. Most of the statements are well known, so that we omit their proofs.

Lemma 10.2 *Let R be defined by (10.1) where the measure μ is quasi-invariant with respect to σ. The following statements hold.*

© Springer International Publishing AG, part of Springer Nature 2018

S. Bezuglyi, P. E. T. Jorgensen, *Transfer Operators, Endomorphisms, and Measurable Partitions*, Lecture Notes in Mathematics 2217,

https://doi.org/10.1007/978-3-319-92417-5_10

(1) R is a positive bounded linear operator with L^1-norm equal to one.

(2) The operator R satisfies the pull-out property: $R[(f \circ \sigma)g] = f R(g)$. Moreover R is a normalized transfer operator if and only if μ is a probability measure.

(3) The operator R can be defined by the following statement: $R(f)$ is a unique element of $L^1(\mu)$ such that, for any function $g \in L^\infty(\mu)$,

$$\int_X g(Rf) \, d\mu = \int_X (g \circ \sigma) f \, d\mu.$$

(4) If μ is σ-invariant, then the operator $S : f \to f \circ \sigma$ is an isometry in $L^p(\mu)$. In this case, the operator R also preserve the measure μ, $\mu R = \mu$.

(5) If μ is a probability σ-invariant measure, then $f \mapsto R(f) \circ \sigma : L^1(X, \mathcal{B}, \mu) \to L^1(X, \sigma^{-1}(\mathcal{B}), \mu)$ is the conditional expectation $\mathbb{E}_\mu(f | \sigma^{-1}(\mathcal{B}))$.

Proof We show only that (5) is true (the other statements are easily verified). For this, we observe that (i) $R(R(f) \circ \sigma) \circ \sigma = R(f) \circ \sigma$ and (ii) for any $\sigma^{-1}(\mathcal{B})$-measurable function g,

$$\int_X g R(f) \circ \sigma \, d\mu = \int_X gf \, d\mu.$$

We calculate the left-hand side integral using σ-invariance of μ and the fact that $g = h \circ \sigma$ for a $\mathcal{B}$-measurable function h:

$$\int_X g R(f) \circ \sigma \, d\mu = \int_X (h \circ \sigma) R(f) \circ \sigma \, d\mu$$
$$= \int_X h R(f) \, d\mu \circ \sigma^{-1}$$
$$= \int_X R((h \circ \sigma)f) \, d\mu$$
$$= \int_X gf \, d\mu$$

$\square$

Suppose now $d\lambda = \varphi d\mu$ where φ is a positive function from $L^\infty(\mu)$. We will find out how the operators R_λ and R_μ are related.

In the above setting, we define the multiplication operator

$$M_\varphi(f) = \varphi f : L^1(\mu) \to L^1(\lambda).$$

Lemma 10.3 *For R_λ, R_μ, φ, and M_φ defined as above, we have*

$$R_\lambda M_\varphi = M_\varphi R_\mu.$$

Proof Indeed, we compute, for a function $f \in L^1(\mu)$,

$$R_\lambda M_\varphi(f) = \frac{(\varphi f d\lambda) \circ \sigma^{-1}}{d\lambda}$$

$$= \frac{(f d\mu) \circ \sigma^{-1}}{d\lambda}$$

$$= \frac{(f d\mu) \circ \sigma^{-1}}{d\mu} \frac{d\mu}{d\lambda}$$

$$= M_\varphi R_\mu(f).$$

$\square$

Let λ be a Borel measure on $(X, \mathcal{B})$ which is equivalent to μ. Then, as we know from Chap. 5, λ is in $\mathcal{L}(R)$. We can find the Radon-Nikodym derivative $W_\lambda = \frac{d(\lambda R_\mu)}{d\lambda}$ of R_μ with respect to λ.

Lemma 10.4 *Let* $\lambda \sim \mu$ *and* $\varphi = \frac{d\lambda}{d\mu}$. *Then* W_λ *is a* σ-*coboundary,* $W_\lambda = (\varphi \circ \sigma)\varphi^{-1}$.

Proof We use the definition of the Radon-Nikodym derivative for the transfer operator and compute

$$\int_X R_\mu(f) \, d\lambda = \int_X R_\mu(f)\varphi \, d\mu$$

$$= \int_X R_\mu(f(\varphi \circ \sigma)) \, d\mu$$

$$= \int_X f(\varphi \circ \sigma) \, d\mu$$

$$= \int_X f(\varphi \circ \sigma)\varphi^{-1} \, d\mu$$

(we used here that $\mu R_\mu = \mu$). The latter means that $W_\lambda = (\varphi \circ \sigma)\varphi^{-1}$. $\square$

Let λ be a quasi-invariant measure with respect to a surjective endomorphism σ of $(X, \mathcal{B})$. Let $\theta_\lambda = \frac{d(\lambda \circ \sigma^{-1})}{d\lambda}$. Then σ generates an operator S on $L^2(\lambda)$ defined by

$$S : f \mapsto f \circ \sigma.$$

It can be seen from Lemma 10.2 (3) that the operators R_λ and S, viewed as operators in $L^2(\lambda)$, form a symmetric pair of operators because

$$\int_X g(R_\lambda f)\, d\lambda = \int_X (Sg)f\, d\lambda, \quad f, g \in L^2(\lambda).$$

So, we can use the notation S^* for R_λ for consistency.

Lemma 10.5 *In the above notation, the operator S^*S is an operator of multiplication M_{θ_λ} by the function θ_λ.*

Proof We note that R_λ is not normalized because

$$R_\lambda(1) = \frac{d(\lambda \circ \sigma^{-1})}{d\lambda} = \theta_\lambda.$$

Then, using inner product in $L^2(\lambda)$, we have

$$\langle S^*S(f), g \rangle_{L^2(\lambda)} = \int_X R_\lambda(f \circ \sigma)g\, d\lambda$$

$$= \int_X f R_\lambda(1)g\, d\lambda$$

$$= \langle M_{\theta_\lambda}(f), g \rangle_{L^2(\lambda)}.$$

The result follows.   □

Theorem 10.6 *Let $(X, \mathcal{B}, \lambda, \sigma)$ be a non-singular dynamical system generated by a surjective endomorphism. Let R_λ be the transfer operator defined by (10.1). The following statements are equivalent:*

(i)   there exists a harmonic function h for R_λ such that h is $\sigma^{-1}(\mathcal{B})$-measurable;
(ii) the Radon-Nikodym derivative θ_λ is a σ-coboundary.

Proof The proof of the theorem is based on Lemma 10.5. We first observe that, since $\lambda \circ \sigma^{-1} \sim \lambda$, the Radon-Nikodym derivative θ_λ is positive a.e. Then the fact that θ_λ is a coboundary, $q\theta_\lambda = q \circ \sigma$, implies that $q \neq 0$.

Therefore, to see that (i) implies (ii), we take a harmonic function for R_λ in the form $h = q \circ \sigma$ and obtain by Lemma 10.5

$$q \circ \sigma = R_\lambda(q \circ \sigma) = (R_\lambda S)(q) = \theta_\lambda q. \tag{10.2}$$

Conversely, if $\theta_\lambda = (q \circ \sigma)q^{-1}$, then

$$R_\lambda(q \circ \sigma) = \theta_\lambda q = q \circ \sigma$$

and the theorem is proved   □

Corollary 10.7 *For the transfer operator* (R_μ, σ) *defined on* $(X, \mathcal{B}, \mu)$, *the measure* $d\lambda = h \, d\mu$ *is* σ-*invariant if and only if* h *is harmonic for* R_μ.

Proof This result follows from the equality where we use Lemma 10.2. For any measurable function $g \in L^\infty(\mu)$, we have

$$\int_X g \, d\lambda = \int_X gh \, d\mu$$

$$= \int_X g R_\mu(h) \, d\mu$$

$$= \int_X (g \circ \sigma) h \, d\mu$$

$$= \int_X g \circ \sigma \, d\lambda.$$

Hence, $\lambda \circ \sigma^{-1} = \lambda$ □

Readers coming from other but related areas, may find the following papers/books useful for background [BLP$^+$10, AR15].

Chapter 11
Piecewise Monotone Maps and the Gauss Endomorphism

Abstract The purpose of the next two chapters is to outline applications of our results to a family of examples of *dynamics of endomorphisms*, and their associated transfer operators.

Keywords Gauss endomorphism · Branching laws · Dynamics if endomorphisms · Piecewise monotone maps

11.1 Transfer Operators for Piecewise Monotone Maps

Earlier papers discussing some of these examples are as follows [Kea72, Lli15, Rad99, Rug16] for the case of *piecewise monotone maps*, [AJL16, BCD16, CL16] for the case of Gauss endomorphism (map), and [Hut81, JMS16, YL16] for the case of iterated function systems. Our emphasis is infinite branching systems.

Readers coming from other but related areas, may find the following papers/books useful for background [AM16, MdF16, JMS16, JLR16, GS16, HŚ16, YZL13, SUZ13, JT15, JPT15].

In this section, we will discuss invariant measures for piecewise monotone maps $\alpha : I \to I$ of an open interval I onto itself. We also consider the corresponding transfer operators (R, α) and show how one can describe R-invariant measures on I. While studying these problems, we assume, for definiteness, that $I = (0, 1)$.

We recall that, by definition, an onto endomorphism α of $(0, 1)$ is called *piecewise monotone* if $(0, 1)$ can be partitioned into a finite or infinite family (J_k) of subintervals $J_k = (t_{k-1}, t_k)$ such that the restriction of α on each J_k is a continuous monotone one-to-one map onto $(0, 1)$ (in many examples, the map α is assumed to be differentiable on each J_k). Since our main interest is focused on invariant non-atomic measures for piecewise monotone maps, we do not need to define α at the

© Springer International Publishing AG, part of Springer Nature 2018
S. Bezuglyi, P. E. T. Jorgensen, *Transfer Operators, Endomorphisms, and Measurable Partitions*, Lecture Notes in Mathematics 2217,
https://doi.org/10.1007/978-3-319-92417-5_11

points of possible discontinuities $\{t_k : k \in \mathbb{N}\}$. In the second part of this section, we apply the proved results to the Gauss map, which is a famous example of a piecewise monotone map. Moreover, since the Gauss map admits a symbolic representation on a product space, we will be able to prove more results about invariant measures for the *Gauss map*.

We notice that the property of piecewise monotonicity of α means that $\alpha : J_k \to (0, 1)$ is a one-to-one map on every interval J_k. Then, for every k, there exists an inverse branch β_k of α such that β_k maps $(0, 1)$ onto J_k and satisfies the condition

$$\alpha \circ \beta_k(x) = x, \qquad x \in (0, 1).$$

We will assume implicitly that the collection of disjoint subintervals (J_k) of $(0, 1)$ is countable.

Let α, $(\beta_k : k \in \mathbb{N})$, and J_k be as above. Suppose that $\pi = (p_k : k \in \mathbb{N})$ is a probability positive vector (probability distribution), i.e., $p_k > 0$ and $\sum_k p_k = 1$.

Definition 11.1 Let a measure μ on $X = (0, 1)$ satisfy the property

$$\mu = \sum_{k=1}^{\infty} p_k \mu \circ \beta_k^{-1}. \tag{11.1}$$

Then μ is called an *iterated function systems measure (IFS measure)* for the iterated function system $(\beta_k : k \in \mathbb{N})$.

It is known that a measure μ satisfying (11.1) is uniquely determined and ergodic, see, e.g., [Hut81]. We will discuss these properties of the measure μ below in Sect. 11.2 and Chap. 12.

The following properties immediately follow from the definitions.

Lemma 11.2

(1) Let μ be an IFS measure for the system $(\beta_k : k \in \mathbb{N})$, defined as in (11.1), where $\beta_k : (0, 1) \to J_k$. Then

$$\mu(J_k) = p_k, \quad k \in \mathbb{N}.$$

(2) For the IFS measure μ and β_k, J_k as above,

$$\mu(A \cap J_k) = \mu(J_k)\mu(\beta_k^{-1}(A)).$$

(3) For $\mu_k := \mu|_{J_k}$, we have $\mu_k \ll \mu$ and $\mu_k \ll \mu \circ \beta_k^{-1}$. Moreover, the Radon-Nikodym derivatives are:

$$\frac{d\mu_k}{d\mu} = \chi_{J_k}, \qquad \frac{d\mu_k}{d(\mu \circ \beta_k^{-1})} = p_k \chi_{J_k}.$$

Proof

(1) Since β_l^{-1} is defined on J_l only, we see that $\mu \circ \beta_l^{-1}(J_k) = 0$ if $l \neq k$. On the other hand, $\mu \circ \beta_k^{-1}(J_k) = 1$ because $\beta_k^{-1}(J_k) = (0, 1)$. Therefore, it follows from (11.1) that $\mu(J_k) = p_k$.

(2) For an IFS measure μ such that $\mu = \sum_{k=1}^{\infty} p_k \mu \circ \beta_k^{-1}$, where $\beta_k : (0, 1) \to J_k$ is a one-to one map and all (J_k) are pairwise disjoint, we find that

$$\mu(A \cap J_k) = \sum_{i=1}^{\infty} p_i \mu \circ \beta_i^{-1}(A \cap J_k)$$

$$= p_k \mu(\beta_k^{-1}(A \cap J_k))$$

$$= p_k \mu(\beta_k^{-1}(A)) \tag{11.2}$$

$$= \mu(J_k)\mu(\beta_k^{-1}(A))$$

(3) By definition, $\mu_k(A) := \mu(A \cap J_k)$. Then $d\mu_k(x) = \chi_{J_k}(x)d\mu(x)$. The other formula in this statement follows from (2) and (11.2). $\qquad\square$

Lemma 11.3 *The IFS measure μ satisfying (11.1) is α-invariant.*

Proof We verify that, for any integrable function f on $X = (0, 1)$,

$$\int_X f \, d(\mu \circ \alpha^{-1}) = \int_X f(\alpha x) \, d\mu$$

$$= \sum_{k=1}^{\infty} p_k \int_X (f \circ \alpha)(x) \, d(\mu \circ \beta_k^{-1})$$

$$= \sum_{k=1}^{\infty} p_k \int_X (f \circ \alpha)(\beta_k x) \, d\mu$$

$$= \int_X f \, d\mu.$$

Hence, $\mu \circ \alpha^{-1} = \mu$. $\qquad\square$

In the next result we answer the following question. Suppose that a piecewise monotone map α is as above, and let (β_k) be the family of inverse branches for α. Let μ be an α-invariant IFS measure of the form (11.1). We address now the following question: how can one determine explicitly the entries of the corresponding probability distribution π in terms of α and the measure μ?

Theorem 11.4 *Let α be a piecewise monotone endomorphism of $(0, 1)$, and let $(J_k : k \in \mathbb{N})$ be the corresponding collection of the disjoint intervals. Suppose that a measure μ is non-atomic and satisfies relation (11.1). Then, the entries (p_k) of the probability distribution $\pi = (p_k : k \in \mathbb{N})$ are determined by formula*

$$p_k = \frac{\int_{J_k} \alpha(x) \, d\mu(x)}{\int_0^1 x \, d\mu(x)} = \mu(J_k). \qquad (11.3)$$

Proof Without loss of generality, we can assume that $\mu(J_k) > 0$ and $J_k \cap J_l = \emptyset$ for all $k \neq l$. Let β_k be the inverse branch of α on the interval J_k. Define the collection of functions $(f_k : k \in \mathbb{N})$ on $(0, 1)$:

$$f_k(x) := \alpha(x) \chi_{J_k}(x), \quad k \in \mathbb{N}. \qquad (11.4)$$

We claim that

$$f_k(\beta_l(x)) = x \delta_{k,l} = \begin{cases} x & \text{if } k = l \\ 0 & \text{if } k \neq l \end{cases}. \qquad (11.5)$$

Indeed, for any $x \in (0, 1)$,

$$f_k(\beta_k(x)) = \alpha(\beta_k(x)) \chi_{J_k}(\beta_k(x)) = x$$

because $\beta_k : (0, 1) \to J_k$. On the other hand, if $l \neq k$, then

$$f_k(\beta_l(x)) = 0,$$

since $\beta_l(x) \in J_l$ and $J_l \cap J_k = \emptyset$.

Next, we notice that if (p_k) is defined according to (11.3), then $\pi = (p_k)$ is a probability distribution:

$$\sum_{k=1}^{\infty} \int_{J_k} \alpha(x) \, d\mu(x) = \int_0^1 \alpha(x) \, d\mu(x) = \int_0^1 x \, d\mu(x)$$

because μ is α-invariant.

To obtain relation (11.3), we first check that $p_l = \mu(J_l)$. Indeed, we can use the fact that $\mu \circ \tau_l^{-k}$, $k \in \mathbb{N}$, is supported by the set J_k and then calculate the measure of J_l by formula (11.1). We get that the right hand side is nonzero only for $k = l$ and $\mu(J_l) = p_l$.

For the other part of (11.3), we calculate

$$\int_{J_k} \alpha(x)\,d\mu(x) = \int_0^1 f_k(x)\,d\mu(x) \qquad \text{(by (11.4))}$$

$$= \sum_l p_l \int_0^1 f_k(\beta_l(x))\,d\mu \qquad \text{(by (11.1))}$$

$$= p_k \int_0^1 f_k \circ \beta_k \,d\mu \qquad \text{(by (11.5))}$$

$$= p_k \int_0^1 x\,d\mu(x),$$

and the result follows. $\square$

The next theorem contains a converse (in some sense) statement for Theorem 11.4.

Theorem 11.5 *Let α be a piecewise monotone map of $(0, 1)$ onto itself. Let $(J_k : k \in \mathbb{N})$ be the collection of open subintervals such that the map α is monotone on each J_k, and let $(\beta_k : (0, 1) \to J_k)$ be the inverse branches for α. Take an α-invariant measure μ on $(0, 1)$, $\mu \circ \alpha^{-1} = \mu$. Suppose that $R = R_\pi$ is the transfer operator acting on measurable functions such that*

$$R(f)(x) = \sum_{k=1}^{\infty} p_k f(\beta_k(x)),$$

where the probability distribution $\pi = (p_k)$ is defined by

$$p_k := \frac{\int_{J_k} \alpha(x)\,d\mu(x)}{\int_0^1 x\,d\mu(x)} = \mu(\beta_k(0, 1)).$$

Then μ is R-invariant if and only if, for any $k, m \in \mathbb{N}$,

$$\left(\int_0^1 x\,d\mu(x)\right)\int_{J_k} \alpha(x)^m\,d\mu(x) = \left(\int_0^1 x^m\,d\mu(x)\right)\int_{J_k} \alpha(x)\,d\mu(x) \qquad (11.6)$$

Proof It was shown in the proof of Theorem 11.4 that the numbers (p_k) defined above satisfy the condition $\sum_k p_k = 1$. For any integers m, k, we set

$$f_{k,m}(x) := \alpha(x)^m \chi_{J_k}(x). \qquad (11.7)$$

Then, for any $k, l, m \in \mathbb{N}$, and $x \in (0, 1)$,

$$f_{k,m}(\beta_l x) = x^m \delta_{k,l}.$$

We apply the following sequence of equivalences to prove the result:

$$\int_0^1 Rf \, d\mu = \int_0^1 f \, d\mu, \qquad \forall f \in \mathcal{F}(X)$$

$$\updownarrow$$

$$\int_0^1 R(f_{k,m}) \, d\mu = \int_0^1 f_{k,m} \, d\mu, \qquad \forall k, m \in \mathbb{N}$$

$$\updownarrow$$

$$p_k \int_0^1 x^m \, d\mu(x) = \int_{J_k} \alpha(x)^m \, d\mu(x), \qquad \forall k, m \in \mathbb{N}.$$

This proves the theorem. $\square$

It follows from Theorem 11.5 that the left hand side of the equality

$$\frac{\int_{J_k} \alpha(x)^m \, d\mu(x)}{\int_0^1 x^m \, d\mu(x)} = \frac{\int_{J_k} \alpha(x) \, d\mu(x)}{\int_0^1 x \, d\mu(x)} = p_k$$

does not depend on m.

Proposition 11.6 *Let α be a piecewise monotone map of $(0, 1)$ onto itself such that $\beta_k : (0, 1) \to J_k$ is an inverse branch for α, $k \in \mathbb{N}$. Suppose μ is an α-invariant measure. The following statements are equivalent:*

(1)

$$\mu = \sum_{k=1}^\infty p_k \mu \circ \beta_k^{-1},$$

(2)

$$(\chi_{J_k} d\mu) \circ \alpha^{-1} = p_k d\mu,$$

(3) $\forall f \in \mathcal{F}((0, 1), \mathcal{B})$,

$$\int_{J_k} f(\alpha x) \, d\mu(x) = p_k \int_0^1 f(x) \, d\mu(x).$$

Proof Let $h(x)$ be a measurable function on $X = (0, 1)$. Then we note that the function $f_k(x) = \chi_{J_k}(x)h(\alpha(x))$ satisfies the relation

$$f_k(\beta_l(x)) = \delta_{k,l}h(x), \qquad \forall k, l \in \mathbb{N}, \tag{11.8}$$

where $\delta_{k,l}$ is the Kronecker delta symbol. Indeed,

$$f_k(\beta_l(x)) = \chi_{J_k}(\beta_l x)h(\alpha(\beta_l(x)))$$

$$= \begin{cases} 0, & \text{if } l \neq k \\ \chi_{J_k}(\beta_k x), & \text{if } k = l \end{cases}$$

$$= \delta_{k,l}h(x).$$

We used the facts that $\beta_k(0, 1) = J_k$, and the sets $(J_k : k \in \mathbb{N})$ are disjoint. Suppose now that (1) holds. Then, for any measurable function φ, we have

$$\int_0^1 \varphi \, d\mu = \sum_{k=1}^{\infty} p_k \int_0^1 \varphi \circ \beta_k \, d\mu.$$

Take $\varphi = \chi_{J_k}(x)h(\alpha(x))$. Hence,

$$\int_0^1 \chi_{J_k}(x)h(\alpha(x)) = \sum_{k=1}^{\infty} p_k \int_0^1 \chi_{J_k}(\beta_k x)h(\alpha(\beta_k x)) \, d\mu(x)$$

$$= p_k \int_0^1 h(x) \, d\mu(x)$$

This relation can be written in the form

$$\int_0^1 h \, (\chi_{J_k} d\mu) \circ \alpha^{-1} = p_k \int_0^1 h \, d\mu \tag{11.9}$$

which is equivalent to statement (2):

$$p_k = \frac{(\chi_{J_k} d\mu) \circ \alpha^{-1}}{d\mu}.$$

Simultaneously, we have shown, in the above proof, that (3) holds due to relation (11.9).

To finish the proof, we observe that all implications are reversible so that the three statements formulated in the proposition are equivalent.

$\square$

Remark 11.7 Given α and μ as above, we can define the transfer operator

$$R\varphi := \frac{(\varphi d\mu) \circ \alpha^{-1}}{d\mu}.$$

Then $R(\mathbf{1}) = \mathbf{1}$ since μ is α-invariant. It follows from Proposition 11.6 that

$$R(\chi_{J_k}) = \frac{(\chi_{J_k} d\mu) \circ \alpha^{-1}}{d\mu} = p_k, \qquad k \in \mathbb{N}.$$

11.2 The Gauss Map

The famous *Gauss endomorphism (map)* σ is an example of a piecewise monotone map with countably many inverse branches. This map has been studied in many papers, we refer, for example, to [CFS82] for basic definitions and facts about σ. Our study below is motivated by [AJL16].

Let $\lfloor x \rfloor$ denote the integer part, and let $\{x\}$ denote the fractional part of $x \in \mathbb{R}$. Then the Gauss map is defined by the formula

$$\sigma(x) = \frac{1}{x} - \left\lfloor \frac{1}{x} \right\rfloor = \left\{ \frac{1}{x} \right\}, \quad 0 < x < 1.$$

We apply the results proved in the first part of this section to this map and find explicit formulas for invariant measures.

The endomorphism σ is a countable-to-one map, which is monotone decreasing on the intervals $J_k = (\frac{1}{k+1}, \frac{1}{k})$. Then the family $\{\tau_k : k \in \mathbb{N}\}$ represents inverse branches for the Gauss map σ:

$$\tau_k(x) = \frac{1}{k+x}, \quad x \in (0, 1).$$

Clearly, τ_k is a monotone decreasing map from $(0, 1)$ onto the subinterval $(\frac{1}{k+1}, \frac{1}{k})$. The relation $(\sigma \circ \tau_k)(x) = x$ holds for any $x \in (0, 1)$, so that τ_k is the inverse branch of σ for every $k \in \mathbb{N}$.

Remark 11.8 We observe that the composition $\tau_{k_1} \circ \cdots \circ \tau_{k_m}$ is also a well defined map from $(0, 1)$ onto the open subinterval $(\tau_{k_1} \circ \cdots \circ \tau_{k_m}(0), \tau_{k_1} \circ \cdots \circ \tau_{k_m}(1))$ of $(0, 1)$ according to the formula:

$$\tau_{k_1} \circ \tau_{k_2} \cdots \circ \tau_{k_m}(x) = \cfrac{1}{k_1 + \cfrac{1}{k_2 + \cdots \cfrac{1}{k_m + x}}} \tag{11.10}$$

It follows from the definition of the Gauss map σ that the endomorphism σ^m is a one-to-one map from each subinterval $(\tau_{k_1} \circ \cdots \circ \tau_{k_m}(0), \ \tau_{k_1} \circ \cdots \circ \tau_{k_m}(1))$ onto $(0, 1)$.

In this section, we are interested in σ-invariant ergodic non-atomic measures. The set of such measures is uncountable. As follows from Lemma 11.3, every IFS measure for σ is σ-invariant (and ergodic as follows from Theorem 11.15).

The following example of an ergodic σ-invariant measure is well known and goes back to Gauss.

Lemma 11.9 (Gauss, [Rén57]) *The class of measures equivalent to the Lebesgue measures dx contains the measure μ_0 with density*

$$ d\mu_0(x) = \frac{1}{\ln 2} \cdot \frac{dx}{(x+1)}, \quad x \in (0, 1). $$

This measure μ_0 is σ-invariant and ergodic.

We apply the methods used in the first part of this section and consider transfer operators R associated to σ, probability distributions $\pi = (p_k : k \in \mathbb{N})$ and the corresponding inverse branches (τ_k).

Lemma 11.10 *Let $\pi = (p_k : k \in \mathbb{N})$ be a probability infinite-dimensional positive vector. Then*

$$ R_\pi f(x) := \sum_{k=1}^{\infty} p_k f(\tau_k x) = \sum_{k=1}^{\infty} p_k f\left(\frac{1}{k+x}\right), \quad x \in (0, 1). \tag{11.11} $$

is a normalized transfer operator associated with the Gauss map σ. Moreover, R_π is normalized, i.e., $R_\pi(1) = 1$.

It is obvious that R_π is positive and normalized. The pull-out property for R_π is verified by direct computations. We omit the details.

Lemma 11.11 *Suppose that λ is a measure on $(0, 1)$ such that $\lambda R_\pi = \lambda$. Then λ is σ-invariant, i.e., $\lambda \circ \sigma^{-1} = \lambda$.*

In fact, this is a particular case of the general statement that holds for any normalized transfer operator. A formal proof was given in Chap. 6.

In what follows we recall a convenient realization of the Gauss map on the space of one-sided infinite sequences. Let

$$ \Omega = \prod_{i=1}^{\infty} \mathbb{N} $$

be the product space with Borel structure generated by cylinder sets

$$ C(k_1, \ldots, k_m) := \{\omega \in \Omega : \omega_1 = k_1, \ldots, \omega_m = k_m\} $$

where $\omega = (a_1, a_2, \ldots)$ denotes an arbitrary point in Ω. Let S be the one-sided shift in Ω:

$$S(a_1, a_2, a_3, \ldots) = (a_2, a_3, \ldots).$$

Clearly, S is a countable-to-one Borel endomorphism of Ω. For every $k \in \mathbb{N}$, we define the inverse branch of S by setting

$$T_k(a_1, a_2, a_3, \ldots) = (k, a_1, a_2, \ldots).$$

Then $T_k(\Omega) = C(k), k \in \mathbb{N}$, and $ST_k = \mathrm{id}_\Omega$. The collection of sets $(C(k) : k \in \mathbb{N})$ forms a partition of Ω.

Take a positive probability distribution $\pi = (p_1, \ldots, p_k, \ldots)$ on the set $\mathbb{N}$ and define the probability product measure

$$\mathbb{P} = \pi \times \pi \times \cdots$$

on Ω so that $\mathbb{P}(C(k_1, \ldots, k_m)) = p_{k_1} \cdots p_{k_m}$. Clearly, the measure $\mathbb{P}$ is S-invariant, i.e., $\mathbb{P} \circ S^{-1} = \mathbb{P}$.

Define the Borel map $F : \Omega \to (0, 1)$ by setting

$$F(\omega) = \cfrac{1}{a_1 + \cfrac{1}{a_2 + \cfrac{1}{a_3 + \cdots}}}$$

where $\omega = (a_1, a_2, a_3, \cdots)$ is any point from Ω.

Remark 11.12 It is clear that F establishes a one-to-one correspondence between sequences from Ω and all irrational points in the interval $(0, 1)$. We denote by X the set $F(\Omega)$. This means that $X = (0, 1) \setminus \mathbb{Q}$. In the sequel, we will use the same notation J_k for the interval $(\frac{1}{k+1}, \frac{1}{k})$ with removed rational points. As was mentioned above, this alternation does not affect continuous σ-invariant measures which are our main object of study.

It is a simple observation that $F(C(k)) = J_k$ for any $k \in \mathbb{N}$. Moreover, it follows from relation (11.10) that

$$F(C(k_1, \cdots, k_m)) = (\tau_{k_1} \circ \cdots \circ \tau_{k_m}(0), \ \tau_{k_1} \circ \cdots \circ \tau_{k_m}(1)). \tag{11.12}$$

The following statement is well known (see e.g. [CFS82]). We formulate it for further references.

Lemma 11.13 *In the above notation, the map F intertwines the pairs of maps σ, S and T_k, τ_k:*

$$F \circ S = \sigma \circ F, \quad F \circ T_k = \tau_k \circ F, \quad k \in \mathbb{N}.$$

From this lemma and relations (11.10) and (11.12), we deduce the following result.

Corollary 11.14 *The collection of intervals $\{(\tau_{k_1} \circ \cdots \circ \tau_{k_m}(0), \ \tau_{k_1} \circ \cdots \circ \tau_{k_m}(1)) : k_1, \ldots k_m \in \mathbb{N}, m \in \mathbb{N}\}$ generates the sigma-algebra of Borel sets on the interval $(0, 1)$.*

Proof This result obviously follows from the facts that the length of subintervals $(\tau_{k_1} \circ \cdots \circ \tau_{k_m}(0), \ \tau_{k_1} \circ \cdots \circ \tau_{k_m}(1))$ tends to zero as $m \to \infty$, and they separate points in $(0, 1)$ because every such subinterval is the image of a cylinder set in Ω.

$\square$

Theorem 11.15 *For any probability distribution $\pi = (p_k : k \in \mathbb{N})$, there is a unique σ-invariant measure μ_π on $(0, 1)$ such that*

$$\int_0^1 f \, d\mu_\pi = \sum_{k=1}^{\infty} p_k \int_0^1 f \circ \tau_k \, d\mu_\pi,$$

or equivalently, μ_π is the IFS measure defined by π and $\{\tau_k\}$ such that

$$\mu_\pi = \sum_{k=1}^{\infty} p_k \mu_\pi \circ \tau_k^{-1}.$$

Conversely, if μ is an IFS measure on $(0, 1)$ with respect to the maps $(\tau_k : k \in \mathbb{N})$, then there exists a product measure $\mathbb{P} = \mathbb{P}_\mu$ on Ω such that $\mu = \mathbb{P} \circ F^{-1}$.

Proof Fix any probability distribution π and consider the stationary product measure

$$\mathbb{P}_\pi = \pi \times \pi \times \cdots$$

on Ω. The measure $\mathbb{P}$ is first determined on cylinder sets, and then it is extended by Kolmogorov consistency to all Borel sets in Ω. In particular, we observe that $\mathbb{P}_\pi(\Omega_k) = p_k$.

Next, we set $\mu_\pi := \mathbb{P}_\pi \circ F^{-1}$, i.e., $\mu_\pi(A) = \mathbb{P}_\pi(F^{-1}(A))$, $\forall A \in \mathcal{B}(0, 1)$. It defines a Borel probability measure on X.

Let φ be any measurable function on Ω. It follows from the above definitions that the following relation holds:

$$\int_{\Omega} \varphi(\omega)\, d\mathbb{P}_{\pi}(\omega) = \sum_{k=1}^{\infty} \int_{\Omega_k} \varphi(\omega)\, d\mathbb{P}_{\pi}(\omega)$$

$$= \sum_{k=1}^{\infty} p_k \int_{\Omega} \varphi(k\omega')\, d\mathbb{P}_{\pi}(\omega').$$

We used in this calculation the fact that $d\mathbb{P}_{\pi}(k\omega') = p_k d\mathbb{P}_{\pi}(\omega')$.

Since F^{-1} is a one-to-one map from X onto Ω, we see that any measurable function f on X is represented as $f = \varphi \circ F^{-1}$. By Lemma 11.13, we obtain that

$$\varphi(k\omega) = \varphi(kF^{-1}x) = f(\tau_k(x)).$$

Therefore, we have

$$\int_0^1 f\, d\mu = \int_{\Omega} \varphi(\omega)\, d\mathbb{P}_{\pi}(\omega)$$

$$= \sum_{k=1}^{\infty} p_k \int_{\Omega} \varphi(k\omega')\, d\mathbb{P}_{\pi}(\omega')$$

$$= \sum_{k=1}^{\infty} p_k \int_0^1 f \circ \tau_k\, d\mu.$$

Hence, this shows that μ_{π} is an IFS measure. By Lemma 11.11, μ_{π} is σ-invariant, and this fact completes the proof.

In order to prove the converse statement, we begin with an IFS measure $\mu = \sum_k p_k \mu \circ \tau_k^{-1}$ and define the product measure $\mathbb{P}$ by setting $\mathbb{P} = \pi \times \pi \times \cdots$ where $\pi = (p_1, p_2, \ldots)$ as in the definition of μ. Then, for any $m \in \mathbb{N}$ and $k_1, \ldots, k_m \in \mathbb{Z}_+$, we have

$$\mathbb{P}(C(k_1, \ldots, k_m)) = \mu(\tau_{k_1} \circ \cdots \circ \tau_{k_m}(0, 1)) = p_1 \cdots p_{k_m}. \tag{11.13}$$

It follows from (11.13) that there is a one-to- one correspondence between IFS measures on $(0, 1)$ and product measures on Ω.

$\square$

We summarize the previous discussions in the following corollary.

Corollary 11.16 *Let σ be the Gauss map, μ a σ-invariant measure. The following are equivalent:*

(i) μ is an IFS measure, $\mu = \sum_{k=1}^{\infty} p_k \mu \circ \tau_k^{-1}$;

(ii) $\mu = \mathbb{P}_\mu \circ F^{-1}$ *for some product measure* $\mathbb{P}$ *on* Ω;
(iii)

$$\mathbb{P}_\mu(C(k_1, \ldots, C_{k_m})) = \mu(\tau_{k_1} \circ \cdots \circ \tau_{k_m}(X))$$

$$= \mu(\tau_{k_1}(0, 1)) \cdots \mu(\tau_{k_m}(0, 1))$$

$$= p_{k_1} \cdots p_{k_m}, \qquad \forall k, m \in \mathbb{N}.$$

Remark 11.17 Suppose that π, π' are two distinct probability distributions. Then the corresponding stationary measures $\mathbb{P}_\pi$ and $\mathbb{P}_{\pi'}$ are mutually singular by the Kakutani theorem [Kak48] for $\pi \neq \pi'$. But it would be interesting to find out whether the map $F : \Omega \to X$ preserve this property. Is it possible to have two σ-invariant measures $\mu_\pi = \mathbb{P}_\pi \circ F^{-1}$ and $\mu_{\pi'} = \mathbb{P}_{\pi'} \circ F^{-1}$ which are both equivalent to the Lebesgue measure?

It turns out that there are σ-invariant measures on $(0, 1)$ which are not generated by product measure on Ω.

Corollary 11.18 *Let* σ *be the Gauss map and let* μ_0 *be the probability* σ-*invariant measure on* $(0, 1)$ *given by the density*

$$d\mu_0(x) = (\ln 2)^{-1} \frac{dx}{x + 1}$$

where dx *is the Lebesgue measure on* $(0, 1)$. *Then* μ_0 *is not an IFS measure.*

Proof We first can directly calculate the measures of the intervals J_k on which the Gauss map σ is one-to-one:

$$\mu_0(J_k) = (\ln 2)^{-1} \int_{(k+1)^{-1}}^{k^{-1}} \frac{1}{x + 1} dx$$

$$= (\ln 2)^{-1} \ln \left(1 + \frac{1}{k(k + 2)}\right). \qquad (11.14)$$

In order to prove the formulated statement, we use Corollary 11.16. Suppose for contrary that μ_0 is an IFS measure. This means that by Theorem 11.15 there exists a product measure $\mathbb{P}_\pi$ such that $\mu_0 = \mathbb{P}_\pi \circ F^{-1}$ for some probability distribution $\pi = (p_1, p_2, \ldots)$. By Corollary 11.16 the measure μ_0 will then satisfy the property

$$\mu_0(\tau_1 \circ \tau_1(0, 1)) = p_1^2 = \mu_0(\tau_1(0, 1))^2. \qquad (11.15)$$

When we calculate the measures of $\tau_1 \circ \tau_1(0, 1)$ and $\tau_1(0, 1)$, we see that

$$\mu_0(\tau_1(0, 1))^2 = \frac{1}{(\ln 2)^2} (\ln(4/3))^2$$

which is not equal to

$$\mu_0(\tau_1 \circ \tau_1(0, 1)) = \frac{1}{\ln 2} \ln(10/9).$$

This is a contradiction that shows that μ_0 is not an IFS measure.   □

Chapter 12
Iterated Function Systems and Transfer Operators

Abstract In this chapter, we will discuss an application of general results about transfer operators, that were proved in previous chapters, to a family of examples based on the notion of *iterated function system (IFS)*.

Keywords Iterated function systems · Potentials · Harmonic functions

12.1 Iterated Function Systems and Measures

We recall that if an endomorphism σ is a finite-to-one map of a a standard Borel space $(X, \mathcal{B})$ onto itself, then there exists a family of one-to-one maps $\{\tau_i\}_{i=1}^n$ such that $\tau_i : X \to X$ and $\sigma \circ \tau_i = \mathrm{id}_X, i = 1, \ldots, n$. The maps τ_i are called the *inverse branches* for σ. The collection of maps $(\tau_i : 1 \leq i \leq N)$ represents an example of iterated function systems. This is a motivating example for the concept of iterated function systems which, in general, not need to have an endomorphism σ but is based on the one-to-one maps $\tau_i : X \to X$ only.

Thus, an IFS consists of a space X and injective maps $\{\tau_i : i \in I\}$ of X into itself. The orbit of any point $x \in X$ is formed by $(\tau_{i_1} \circ \cdots \circ \tau_{i_k}(x) : i_1, \ldots, i_k \in I, k \in \mathbb{N})$. The study of properties of an IFS assumes that the underlying space X is a complete metric space (or compact space) and the maps τ_i are continuous (or even contractions).

The main tool in our study of an IFS $(\tau_i : i = 1, \ldots, n)$ on a space X is a realization of the IFS as the full one-sided shift S on a symbolic product space Ω. Then any S-invariant and ergodic product-measure $\mathbb{P}$ on Ω can be pulled back to X. This construction gives ergodic invariant measures for IFSs.

© Springer International Publishing AG, part of Springer Nature 2018
S. Bezuglyi, P. E. T. Jorgensen, *Transfer Operators, Endomorphisms,
and Measurable Partitions*, Lecture Notes in Mathematics 2217,
https://doi.org/10.1007/978-3-319-92417-5_12

Remark 12.1 We discuss here the case of *finite* iterated function systems only. The theory of *infinite* iterated function systems is more complicated and require additional assumptions (see, for example, the expository article [Mau95], [Hut96], and [HR00]). It worth recalling that we have already dealt with infinite IFS in Chap. 11 when we discussed piecewise monotone maps and the Gauss map.

Let $p = (p_i : i = 1, \ldots, N)$ be a strictly positive probability vector, i.e., $\sum_{i=1}^{N} p_i = 1$ and $p_i > 0$ for all i. A measure μ_p on a Borel space $(X, \mathcal{B})$ is called an *IFS measure* for the iterated function system $(\tau_i : i = 1, \ldots, N)$ if

$$\mu_p = \sum_{i=1}^{N} p_i \, \mu_p \circ \tau_i^{-1}, \tag{12.1}$$

or, equivalently,

$$\int_X f(x) \, d\mu_p(x) = \sum_{i=1}^{N} p_i \int_X f(\tau_i(x)) \, d\mu_p(x), \qquad f \in L^1(\mu_p).$$

In particular, p can be the uniformly distributed probability vector, $p_i = 1/N$. Then the corresponding measure ν_p satisfies the property

$$\nu_p = N^{-1} \sum_{i=1}^{N} \nu_p \circ \tau_i^{-1}.$$

Definition 12.2 Let $(X; \tau_1, \ldots, \tau_n)$ be a finite iterated function system, and let $p = (p_i)$ be a positive probability vector. Define a positive linear operator acting in the space of Borel functions:

$$R(f)(x) = \sum_{i=1}^{N} p_i W(\tau_i x) f(\tau_i x) \tag{12.2}$$

where W is a nonnegative Borel function (sometimes it is called a *weight*). If, for all x, one has

$$\sum_{i=1}^{N} p_i W(\tau_i x) = 1,$$

then R is a normalized transfer operator in the sense that $R(\mathbf{1}) = \mathbf{1}$.

To clarify our terminology, we note that R is not, in general, a transfer operator because the maps $(\tau_1, \ldots, \tau_N)$ do not define an endomorphism σ. But if the IFS $(X; \tau_1, \ldots, \tau_N)$ consists of inverse branches for a finite-to-one onto endomorphism σ, then the operator R is a transfer operator related to σ: for $y_i = \tau_i x$, we have

$$R(f)(x) = \sum_{y_i : \sigma(y_i) = x} p_{y_i} W(y_i) f(y_i).$$

In some cases, it is convenient to modify W by considering $\widetilde{W}(y) = p_y W(y)$.

It is not difficult to see that if $(X; \tau_1, \ldots, \tau_N)$ is an IFS, then the maps (τ_i) are the inverse branches for an endomorphism σ if and only if the sets $J_i = \tau_i(X)$ have the properties:

$$X = \bigcup_{i=1}^{N} J_i, \qquad J_i \cap J_k = \emptyset, \ i \neq k.$$

Indeed, one can then define $\sigma(x) = \tau_i^{-1}(x), x \in J_i, 1 \leq i \leq N$.

If an IFS is generated by inverse branches of an endomorphism σ, then we can add more useful relations. For any Borel set $A \subset X$, we see that

$$\sigma^{-1}(A) = \bigcup_i \tau_i(A)$$

and, more generally,

$$\sigma^{-k}(A) = \bigcup_{\omega|k} \tau_{\omega|k}(A).$$

As for abstract transfer operators, we define the notion of integrability of R: we say that R is integrable with respect to a measure ν if $R(1) \in L^1(\nu)$. In this case, R acts on the measure ν, $\nu \mapsto \nu R$. Then we can define the set $\mathcal{L}(R)$ of all measures on $(X, \mathcal{B})$ such that $\nu P \ll \nu$.

Lemma 12.3 *Let $(X; \tau_1, \ldots, \tau_N)$ be an IFS, and let $p = (p_i)$ be a probability distribution on $\{1, \ldots, N\}$. Suppose that R is the operator defined for the IFS by (12.2), and the measure μ_p satisfies (12.1). Then R is μ_p-integrable if and only if $W \in L^1(\mu_p)$. Furthermore, μ_p belongs to $\mathcal{L}(R)$ and*

$$\frac{d\mu_p R}{d\mu_p} = W.$$

Proof We use (12.2) and show that

$$\int_X R(f)(x)\,d\mu_p(x) = \int_X \sum_{i=1}^N p_i\, W(\tau_i x) f(\tau_i x)\,d\mu_p(x)$$

$$= \int_X f W\Big(\sum_{i=1}^N p_i\,d\mu_p \circ \tau_i^{-1}\Big)$$

$$= \int_X f W\,d\mu_p.$$

This calculation shows that the following facts hold. Firstly, $R(1)$ is μ_p-integrable if and only if $W \in L^1(\mu_p)$; secondly, $\mu_p \in \mathcal{L}(R)$; thirdly, the Radon-Nikodym derivative of $\mu_p R$ with respect to μ_p is W.   □

The question about the existence of an IFS measure for a given finite or infinite iterated function system $(\tau_i : i \in I)$ is of extreme importance. We discuss here a general scheme that leads to IFS measures. We do not formulate rigorous statements; instead we describe the construction method. This approach works perfectly for many specific applications under some additional conditions on X and maps τ_i. For definiteness, we assume that $I = \{1, \ldots, N\}$.

Let Ω be the product space:

$$\Omega = \prod_{i=1}^{\infty} \{1, \ldots, N\}.$$

Our goal is to define a map F (a coding map) from Ω to X. In general, $F(\Omega)$ will be a subsets of X called the attractor of the IFS.

For any infinite sequence $\omega = (\omega_1, \omega_2, \ldots) \in \Omega$, let $\omega|_n$ denote the finite truncation, i.e., $\omega|_n$ is the finite word $(\omega_1, \ldots, \omega_n)$. Then, we can use this word $\omega|_n$ to define a map $\tau_{\omega|_n}$ acting on X by the formula:

$$\tau_{\omega|_n}(x) := \tau_{\omega_1} \cdots \tau_{\omega_n}(x), \qquad x \in X, \; \omega \in \Omega, \; n \in \mathbb{N}.$$

It is said that Ω is an *encoding space* if, for every $\omega \in \Omega$,

$$F(\omega) = \bigcap_{n \geq 1} \tau_{\omega|_n}(X) \tag{12.3}$$

is a singleton. In other words, we have a well defined Borel map F from Ω to $F(\Omega)$ where $x = F(\omega)$ is defined by (12.3). It is worth noting that there are IFS such that $F(\Omega) = X$. One of such IFS will be discussed in this section below.

There are various sufficient conditions under which there exists a *coding map* $F : \Omega \to X$ for a given IFS. For instance, this is the case when each τ_i is a contraction and X is a complete metric space.

Next, we define the following maps on Ω: the left shift $\tilde{\sigma}$ by setting

$$\tilde{\sigma}(\omega_1, \omega_2, \ldots) = (\omega_2, \omega_3, \ldots),$$

and the inverse branches $\tilde{\tau}_i$ of $\tilde{\sigma}$ by setting

$$\tilde{\tau}_i(\omega_1, \omega_2, \ldots) = (i, \omega_1, \omega_2, \ldots), \quad i = 1, \ldots, N.$$

Clearly,

$$\tilde{\tau}_i(\Omega) = C(i) = \{\omega \in \Omega : \omega_1 = i\},$$

and the space Ω is partitioned by the sets $C(i), i = 1, \ldots, N$.

The following statement follows directly from the definitions.

Lemma 12.4 *The map* $F : \Omega \to X$ *is a factor map, i.e.,*

$$F \circ \tilde{\tau}_i = \tau_i \circ F, \qquad i = 1, \ldots, N.$$

Let $p = (p_i : i = 1, \ldots, N)$ be a positive probability vector. It defines the product measure $\mathbb{P}$ on Ω,

$$\mathbb{P} = p \times p \times \cdots.$$

Firstly, $\mathbb{P}$ is defined on the algebra of cylinder sets $(C(i_1, \ldots, i_m) : 1 \le i_1, \ldots, i_m \le N, m \in \mathbb{N})$ by the formula

$$\mathbb{P}(C(i_1, \ldots, i_m)) = p_{i_1} \cdots p_{i_m},$$

and then $\mathbb{P}$ is extended to the sigma-algebra of Borel sets on Ω by the standard procedure.

We observe that the maps $(\tilde{\tau}_1, \ldots, \tilde{\tau}_N)$ constitute an IFS on Ω such that $\mathbb{P}$ is an IFS measure:

$$\mathbb{P} = \sum_{i=1}^{N} p_i \, \mathbb{P} \circ \tilde{\tau}_i^{-1}. \tag{12.4}$$

Proposition 12.5 *Suppose that* $(X; \tau_1, \ldots, \tau_n)$ *is an IFS that admits a coding map* $F : \Omega \to X$. *Let* $p = (p_i)$ *be a probability vector generating the product measure* $\mathbb{P} = p \times p \times \cdots$. *Then the measure*

$$\mu := \mathbb{P} \circ F^{-1}$$

is an IFS measure satisfying

$$\mu = \sum_{i=1}^{N} p_i \mu \circ \tau_i^{-1}.$$

Moreover, if F is continuous, then μ has full support.

Proof By definition of measure μ_p, we have

$$\mu_p(A) = \mathbb{P}(F^{-1}(A)).$$

Then we use Lemma 12.4 and (12.4) to show that μ_p is an IFS measure:

$$\mu_p(A) = \mathbb{P}(F^{-1}(A))$$

$$= \sum_{i=1}^{N} p_i \; \mathbb{P}(\tilde{\tau}_i(F^{-1}(A)))$$

$$= \sum_{i=1}^{N} p_i \; \mathbb{P}(F^{-1}(\tau_i(A)))$$

$$= \sum_{i=1}^{N} p_i \; \mu_p(\tau_i(A)).$$

Let now C be an open subset of X. Then there exists a cylinder set $C(\omega_1, \ldots, \omega_m)$ such that

$$\tilde{\tau}_{\omega_1} \circ \cdots \circ \tilde{\tau}_{\omega_m}(\Omega) \subset F^{-1}(C).$$

It follows from this inclusion that

$$\mu_p(C) = \mathbb{P}(F^{-1}(C)) \geq \mathbb{P}(C(\omega_1, \ldots, \omega_m)) = p_{\omega_1} \cdots p_{\omega_m}.$$

The proof is complete.   □

12.2   Transfer Operator for $x \mapsto 2x \bmod 1$

We consider here the one of the most popular endomorphisms, $\sigma : x \mapsto 2x$ mod 1 defined on the unit interval $[0, 1]$ and study its properties related to the corresponding transfer operator and iterated function system.

We fix the following notations for this section. Let $X = [0, 1] = \mathbb{R}/\mathbb{Z}$, $\sigma(x) = 2x \bmod 1$,

$$\tau_0(x) = \frac{x}{2}, \quad \tau_1(x) = \frac{x+1}{2}, \quad x \in X,$$

and let $\lambda = dx$ denote the Lebesgue measure on X. Then $(X; \tau_0, \tau_1)$ is an IFS defined by the inverse branches for σ.

In this example, we will illustrate the facts about transfer operators (R, σ) by considering specific weights W.

We use the formula given in (12.2) to define the transfer operator by a weight function W.

Take the function $W = \cos^2(\pi x)$. Define the transfer operator associated with the IFS $(X; \tau_0, \tau_1)$:

$$R_W(f)(x) = \cos^2(\frac{\pi x}{2}) f(\frac{x}{2}) + \cos^2(\frac{\pi(x+1)}{2}) f(\frac{x+1}{2}).$$

Since $\cos^2(\frac{\pi(x+1)}{2}) = \sin^2(\frac{\pi x}{2})$, we get that

$$R_W(f)(x) = \cos^2(\frac{\pi x}{2}) f(\frac{x}{2}) + \sin^2(\frac{\pi x}{2}) f(\frac{x+1}{2}). \tag{12.5}$$

It follows from (12.5) that the transfer operator R_W is normalized because

$$R_W(1) = \cos^2(\frac{\pi x}{2}) + \sin^2(\frac{\pi x}{2}) = 1.$$

Lemma 12.6 *The Lebesgue measure $\lambda = dx$ belongs to $\mathcal{L}(R_p)$ for any probability vector p. If $p_0 = p_1 = 1/2$, then*

$$\frac{d\lambda R_{1/2}}{d\lambda}(x) = \cos^2(\pi x).$$

Proof The fact that $\lambda R_p \ll \lambda$ will be clear from the following computation which are conducted for the case $p_0 = p_1 = 1/2$:

$$\int_X R(f)\, dx = 2^{-1} \int_X \cos^2(\frac{\pi x}{2}) f(\frac{x}{2})\, dx + 2^{-1} \int_X \sin^2(\frac{\pi x}{2}) f(\frac{x+1}{2})\, dx$$

$$= \int_0^{1/2} \cos^2(\pi y) f(y)\, dy + \int_{1/2}^1 \cos^2(\pi y) f(y)\, dy$$

$$= \int_X \cos^2(\pi y) f(y)\, dy$$

This means that $2W = \cos^2(\pi x)$ is the Radon-Nikodym derivative. $\square$

In what follows we will use the formula

$$R(f)(x) = \sum_{y:\sigma y=x} \cos^2(\pi y) f(y) \tag{12.6}$$

for the transfer operator. The advantage of this definition is that R is now normalized, $R(1) = 1$. It follows from Lemma 12.6 that

$$d(\lambda R) = 2\cos^2(\pi x) d\lambda. \tag{12.7}$$

Let δ_0 denote the atomic Dirac measure concentrated at $x = 0$.

Corollary 12.7

(1) The measures δ_0 and λ are σ-invariant.
(2) The measures δ_0 and λ are R-invariant, λR is absolutely continuous with respect to λ but $\delta_0 \notin \mathcal{L}(R)$.

Proof The first statement is obvious and well known.

To show that (2) holds, we notice that $R(f)(0) = f(0)$, and this fact can be interpreted as

$$\int_X f \, d(\delta_0 R) = \int_X f \, d\delta_0.$$

Hence δ_0 is R-invariant. Since $\delta_0 R = 1/2(\delta_0 + \delta_{1/2})$ it is clear that $\delta_0 R$ is not absolutely continuous with respect to δ_0. The result about the Lebesgue measure λ has been proved in Lemma 12.6.

□

We recall that, for given R, σ, one can define linear operators $\widehat{S}$ and $\widehat{R}$ in the universal Hilbert space $\mathcal{H}(X)$. By definition (see Chap. 8),

$$\widehat{S}(f\sqrt{d\mu}) = (f \circ \sigma)\sqrt{d(\lambda R)}.$$

Proposition 12.8 *The operator $\widehat{S}$ is an isometry in the Hilbert space $L^2(\lambda)$ where λ is the Lebesgue measure on $X = [0, 1]$.*

Proof We first note that the following useful formula holds:

$$\int_X f(\sigma x)g(x) \, dx = \int_X f(x)\frac{g(\tau_0 x) + g(\tau_1 x)}{2} \, dx. \tag{12.8}$$

Then we use (12.6) and (12.7) to find the norm

$$\|\widehat{S}(f\sqrt{d\lambda})\|^2_{L^2(\lambda)} = \int_X f(2x \bmod 1)^2 2\cos^2(\pi x)\, d\lambda(x)$$

$$= \int_X f(x)^2 (\cos^2(\pi x/2) + \sin^2(\pi x/2))\, dx$$

$$= \|f\,d\lambda\|^2_{L^2(\lambda)}.$$

Hence $\widehat{S}$ is an isometry.   □

We recall that the operator $\widehat{R}(f\sqrt{d\mu}) = (Rf)\sqrt{d(\mu \circ \sigma^{-1})}$ is the adjoint operator $\widehat{S}^*$ for $\widehat{S}$. For the Lebesgue measure λ, one has $\lambda \circ \sigma^{-1} = \lambda$, hence

$$\widehat{R}(f\sqrt{d\lambda}) = (Rf)\sqrt{d\lambda}.$$

Corollary 12.9 *For the Lebesgue measure λ on $X = [0, 1]$, the projection $\widehat{E} = \widehat{S}\widehat{R}$ from $\mathcal{H}(X)$ onto $\mathcal{H}(\lambda)$ acts by the formula*

$$\widehat{E} = \sqrt{2}(R(f) \circ \sigma)|\cos(\pi x)|\sqrt{d\lambda},$$

where $R(f) \circ \sigma = \mathbb{E}_\lambda(\cdot \mid \sigma^{-1}(\mathcal{B}))$ is the conditional expectation.

Example 12.10 In this example, we give several formulas for the action of the transfer operator R on any Dirac measure δ_a.

For $a = 0$, we observe that

$$\widehat{S}(f\sqrt{d\delta_0}) = (f \circ \sigma)\sqrt{d\delta_0}, \qquad \widehat{R}(f\sqrt{d\delta_0}) = R(f)\sqrt{d\delta_0}$$

because δ_0 is simultaneously R-invariant and σ-invariant.

By direct computation we find that, for $a \in (0, 1)$,

$$\widehat{S}(f\sqrt{d\delta_a}) = (f \circ \sigma)\sqrt{\cos^2(\pi a/2)\delta_a + \sin^2(\pi a/2)\delta_{(a+1)/2}},$$

$$\widehat{R}(f\sqrt{d\delta_a}) = R(f)\sqrt{d\delta_a},$$

and

$$\widehat{S}\widehat{R}(f\sqrt{d\delta_a}) = R(f) \circ \sigma\sqrt{\cos^2(\pi a)\delta_a + \sin^2(\pi a)\delta_{a+1/2}}.$$

In particular, if $a = 1/2$, we can show more.

Lemma 12.11 *For the Dirac measure $\delta_{1/2}$, the following relations hold:*

$$(\delta_{1/2} \circ \sigma^{-1})R = \delta_0,$$

$$\widehat{E}(f\sqrt{d\delta_{1/2}}) = f(0)\sqrt{d\delta_0}$$

where $\widehat{E} = \widehat{S}\widehat{R}$.

Proof These results follow from the above formulas by straightforward computations.   □

Chapter 13
Examples

Abstract In this chapter, we discuss in detail several examples of transfer operators that are mentioned in Introduction.

Keywords Endomorphisms · Transfer operators · Measurable partitions · Markov processes

13.1 Transfer Operator and a System of Conditional Measures

For every measurable partition ξ of a probability measure space $(X, \mathcal{B}, \mu)$, there exists a system of conditional measures [Roh49b]. We apply this remarkable result to the case of a surjective endomorphism.

Theorem 13.1 *Let $(X, \mathcal{B}, \mu)$ be a standard measure space with finite measure, and let σ be a surjective homeomorphism on X. Let ξ be the measurable partition into pre-images of σ, $\xi = \{\sigma^{-1}(x) : x \in X\}$. Then there exists a system of conditional measures $\{\mu_C\}_{C \in \xi}$ defined uniquely by μ and ξ, see Definition 2.7. For an onto endomorphism, X/ξ is identified with X, and the following disintegration formula holds:*

$$\int_X f(x)\, d\mu(x) = \int_X \left(\int_{C_x} f(y)\, d\mu_{C_x}(y) \right) d\mu_\sigma(x)$$

where μ_σ is the restriction of μ to $\sigma^{-1}(\mathcal{B})$, and x is identified with C_x.

© Springer International Publishing AG, part of Springer Nature 2018
S. Bezuglyi, P. E. T. Jorgensen, *Transfer Operators, Endomorphisms, and Measurable Partitions*, Lecture Notes in Mathematics 2217, https://doi.org/10.1007/978-3-319-92417-5_13

In Example 1.6, we introduced a transfer operator (R, σ) on a standard probability measure space $(X, \mathcal{B}, \mu)$. Here we consider a slightly more general construction by setting

$$Rw(f)(x) := \int_{C_x} f(y)W(y)\, d\mu_{C_x}(y) \tag{13.1}$$

where C_x is the element of ξ containing x and W is a *positive* μ-integrable function. It follows then that f is μ_{C_x}-integrable functions for a.e. x. We note first that the condition $\mu(X) = 1$ implies that $\mu_C(X) = 1$ for a.e. $C \in X/\xi$. Another important fact is that the quotient measure space defined by ξ is isomorphic to $(X, \sigma^{-1}(\mathcal{B}), \mu_\sigma)$ (see Sect. 2.3 for details).

It is natural to consider the operator R_W acting either in $L^1(X, \mathcal{B}, \mu)$ or in $L^2(X, \mathcal{B}, \mu)$ in this example.

Proposition 13.2

(1) If σ is an onto endomorphism of a probability measure space $(X, \mathcal{B}, \mu)$, then (R_W, σ) is a transfer operator on $(X, \mathcal{B}, \mu)$. It is normalized if and only if

$$\int_{C_x} W\, d\mu_{C_x} = 1$$

for μ-a.e. $x \in X$.
(2) For any measurable function f, the function $R_W(f)(x)$ is constant a.e. on every C_x for μ-a.e. x.

Proof

(1) Clearly, R_W is positive. To see that the pull-out property holds, we calculate

$$R_W((f \circ \sigma)g)(x) = \int_{C_x} (f \circ \sigma)(y)g(y)W(y)\, d\mu_{C_x}$$

$$= f(x)\int_{C_x} g(y)W(y)\, d\mu_{C_x}(y)$$

$$= f(x)R_W(g)(x).$$

Here we used the fact that $f(\sigma(y)) = f(x)$ for $y \in C_x = \sigma^{-1}(x)$.
We also obtain from (13.1) that R_W is normalized if and only if

$$R_W(1)(x) = \int_{C_x} W(y)\, d\mu_{C_x}(y) = 1$$

for any $x \in X$.

(2) Because the value of $R_W(f)$ evaluated at x is the integral of f over the measure space (C_x, μ_x), we see that the value of $R_W(f)$ at $x' \in C_x$ is equal to $R_W(f)(x)$. Hence, the function $R_W(f)$ is constant a.e. on every element C_x of the partition ξ.

$\square$

Proposition 13.2 allows us to deduce several simple consequences of the proved results.

Corollary 13.3

(1) For the transfer operator (R_W, σ), the following property holds a.e.

$$R_W^2(f)(x) = R_W(f)(x)R(1)(x).$$

(2) If $\int_{C_x} W \, d\mu_{C_x} = 1$ for a.e. x, then, for any f, $R_W(f)$ is a harmonic function with respect to R_W.

(3) For any $\mathcal{B}$-measurable function f, the function $R(f)$ is $\sigma^{-1}(\mathcal{B})$-measurable.

In particular, the case when $W = 1$, gives a simple straightforward example of harmonic functions for the corresponding transfer operator R_1.

Proof All these results follow directly from the fact that $R_W(f)$ is constant on elements C of the partition ξ. $\square$

Proposition 13.4 *For the transfer operator R_W defined by (13.1),*

$$\int_X R_W(f)(x) \, d\mu(x) = \int_X f(x)W(x) \, d\mu(x), \qquad f \in L^1(\mu),$$

that is $W = \dfrac{d\mu R_W}{d\mu}$. If $W = 1$, then R is an isometry in the space $L^1(X, \mathcal{B}, \mu)$.

Proof The proof follows from the following calculations based on (2.3):

$$\int_X R_W(f)(x) \, d\mu(x) = \int_X \left(\int_{C_x} f(y)W(y) \, d\mu_{C_x}(y) \right) d\mu(x)$$

$$= \int_X \int_C \left(\int_{C_x} f(y)W(y) \, d\mu_{C_x}(y) \right) d\mu_C(z) \, d\mu_\sigma(C)$$

$$= \int_X \int_{C_x} \left(\int_C f(y)W(y) \, d\mu_C(z) \right) d\mu_{C_x}(y) \, d\mu_\sigma(C)$$

$$= \int_X \left(\int_{C_x} f(y)W(y) \, d\mu_{C_x}(y) \right) d\mu_\sigma(C)$$

$$= \int_X f(x)W(x) \, d\mu(x)$$

We used here that μ_C is a probability measure for a.e. C. This equality shows that $d(\mu R_W) = d\mu$.

$\square$

Let σ be an endomorphism of a standard Borel space $(X, \mathcal{B})$. Suppose now that ξ is a σ-invariant partition of $(X, \mathcal{B}, \mu)$. This means that $\sigma^{-1}(C)$ is a ξ-set for any element C of the partition ξ. Take a measure ν on $(X/\xi, \mathcal{B}/\xi)$. Denote by $(\nu_C)_{C \in X/\xi}$ a random measure on $(X, \mathcal{B})$, i.e., it satisfies the conditions: (i) $C \to \nu_C(B)$ is measurable for any $B \in \mathcal{B}$, (ii) $\nu_C(C) \in L^1(X/\xi, \nu)$. We *define* a measure μ on $(X, \mathcal{B})$ by setting

$$\mu(B) = \int_{X/\xi} \nu_C(B) \, d\nu(C). \tag{13.2}$$

Corollary 13.5 *Suppose that the measure* μ *on* $(X, \mathcal{B})$ *is as in (13.2). Let the transfer operator* R *be defined by the relation*

$$R(f)(x) := \int_{C_x} f(y) \, d\nu_{C_x}(y).$$

Then the Radon-Nikodym derivative $W = \dfrac{d\mu R}{d\mu} = \nu_C(C).$

The *proof* is the same as in Proposition 13.4.

For the class of transfer operators which are considered in this example, we can easily point out harmonic functions.

Theorem 13.6 *Let* σ *be an onto endomorphism of a probability standard measure space* $(X, \mathcal{B}, \mu)$, *and* ξ *is a* σ-*invariant measurable partition of* X. *Define a transfer operator* R *by setting*

$$R(f)(x) = \int_{C_x} f(y) \, d\mu_{C_x}(y)$$

where (μ_{C_x}) *is the system of conditional measures associated to* ξ. *Then a measurable function* h *defined on* X *is harmonic with respect to* R *if and only if* h *is* ξ-*measurable, i.e.,* $h(x)$ *is constant on every element* C *of* ξ.

Proof We note that μ_C is a probability measure for $C \in X/\xi$. Therefore, if $h(x) = h(C_x)$ for all $x \in X$, then

$$R(h)(x) = \int_{C_x} h(y) \, d\mu_{C_x}(y) = h(C_x)\mu_{C_x}(C_x) = h(x).$$

Conversely, if for all $x \in X$, we have $R(h)(x) = h(x)$, then $h(x)$ satisfies the relation

$$h(x_1) = \int_C h(y) \, d\mu_C(y) = h(x_2)$$

where x_1 and x_2 are taken from C. $\square$

Example 13.7 Consider a standard probability measure space $(X, \mathcal{B}, \mu)$, and let $\nu : x \mapsto \nu_x$, $x \in X$, be a random measure taken values in $M(Y)$ where $(Y, \mathcal{A})$ is a measurable space. Consider the product measure space $(X \times Y, m)$ where

$$m = \int_X \nu_x \, d\mu$$

is a measure on $\mathcal{B} \times \mathcal{A}$.

Define an operator $R : \mathcal{F}(X \times Y) \to \mathcal{F}(X)$:

$$R(f)(x) = \int_X f(x, y) K(x, y) \, d\nu_x(y) \tag{13.3}$$

where $K(x, y)$ is a non-negative measurable bounded function.

Let $\mathcal{F}_X$ be the set of functions depending on $x \in X$ only. In other words, $\mathcal{F}_X = \mathcal{F}(X \times Y) \circ \pi$ where $\pi : X \times Y \to X$ is the projection.

Claim. The operator R satisfies the property:

$$R(fg) = f R(g)$$

if $f \in \mathcal{F}_X$ and $g \in \mathcal{F}(X \times Y)$.

Indeed, we see that

$$R((f \circ \pi)g)(x) = \int_X K(x, y) f(\pi(x, y)) g(x, y) \, d\nu_x(y)$$

$$= f(x) \int_X K(x, y) g(x, y) \, d\nu_x(y)$$

$$= f(x) R(g)(x).$$

Example 13.8 In this example, we will work with a countable-to-one (or bounded-to-one) endomorphism σ of a probability measure space $(X, \mathcal{B}, \mu)$. As was mentioned in Theorem 2.10, there is a partition $(A_i | i \in I)$ of X into measurable sets of positive measure such that $\sigma \circ \tau_i = \mathrm{id}$ on A_i.

We define a transfer operator $R_{\tau_i} : \mathcal{M}(\sigma(A_i)) \to \mathcal{M}(\sigma(A_i))$ by setting

$$(R_i f)(x) = f(\tau_i(x)), \quad x \in \sigma(A_i). \tag{13.4}$$

Then R_{τ_i} is positive and $R_{\tau_i}((f \circ \sigma)g) = f \circ (\sigma \tau_i)g \circ \tau_i = f R_{\tau_i}(g)$.

This example can be discussed in detail in the context of stationary *Bratteli diagrams*, see [BK16, BJ15] for more details. The notion of Bratteli diagrams is widely used for constructions of transformation models in various dynamics. It is difficult to overestimate the significance of Bratteli diagrams for the study of dynamical systems. A class of graduated infinite graphs, later called Bratteli diagrams, was originally introduced by Bratteli [Bra72] in his breakthrough article on the classification of approximately finite (AF) C^*-algebras. It turned out that the close ideas developed by A. Vershik in the study of sequences of measurable partitions led to a realization of any ergodic automorphism of a standard measure as a transformation acting on a path space of a graph (afterwards called a *Vershik map*) [Ver81].

Example 13.9 (Parry's Jacobian and Transfer Operator [PW72]) Let σ be a bounded-to-one nonsingular endomorphism of $(X, \mathcal{B}, \mu)$, and let $(A_i | i \in I)$ be the corresponding Rohlin partition.

We define for $x \in A_i$

$$J_i(x) = \frac{d\mu\tau_i}{d\mu}(x), \quad x \in A_i$$

and let

$$J(x) = \sum_{i \in I} J_i(x)\chi_{A_i}(x), \quad x \in X. \tag{13.5}$$

The function $J(x) = J_\sigma(x)$ is called *Jacobian* and was defined in [Par69], [PW72]. One can prove that the function $J(x)$ is independent of the choice of the Rohlin partition.

By non-singularity of σ we have the following relations

$$\theta_\sigma(x) := \frac{d\mu\sigma^{-1}}{d\mu}(x) = \sum_{y \in \sigma^{-1}(x)} \frac{1}{J(x)},$$

$$\omega_\sigma(x) := \frac{d\mu}{d\mu\sigma^{-1}}(x) = \frac{1}{\theta_\sigma(\sigma x)}.$$

The function ω_σ is called the *Radon-Nikodym derivative of* σ and satisfies the property

$$\int_X f \circ \sigma \, \omega_\sigma \, d\mu = \int_X f \, d\mu$$

for all $f \in L^1(X, \mu)$.

We use the Jacobian to define the transfer operator R_σ acting on $\mathcal{M}(X)$:

$$(R_\sigma h)(x) = \sum_{y \in \sigma^{-1}(x)} \frac{h(y)}{J_\sigma(y)}.$$

It can be easily verified that R_σ satisfies the characteristic property for transfer operators.

References

[AA01]   Y.A. Abramovich, C.D. Aliprantis, Positive operators, in *Handbook of the Geometry of Banach Spaces*, vol. I (North-Holland, Amsterdam, 2001), pp. 85–122

[AU15]   S. Albeverio, S. Ugolini, A Doob h-transform of the Gross-Pitaevskii Hamiltonian. J. Stat. Phys. **161**(2), 486–508 (2015)

[AGZ15] L. Alili, P. Graczyk, T. Żak, On inversions and Doob *h*-transforms of linear diffusions, in *In Memoriam Marc Yor—Séminaire de Probabilités XLVII*, vol. 2137. Lecture Notes in Mathematics (Springer, Cham, 2015), pp. 107–126

[AJ15]   D. Alpay, P.E.T. Jorgensen, Spectral theory for Gaussian processes: reproducing kernels, boundaries, and L^2-wavelet generators with fractional scales. Numer. Funct. Anal. Optim. **36**(10), 1239–1285 (2015)

[AK13]   D. Alpay, A. Kipnis, A generalized white noise space approach to stochastic integration for a class of Gaussian stationary increment processes. Opusc. Math. **33**(3), 395–417 (2013)

[AK15]   D. Alpay, A. Kipnis, Wiener chaos approach to optimal prediction. Numer. Funct. Anal. Optim. **36**(10), 1286–1306 (2015)

[AL13]   D. Alpay, I. Lewkowicz, Convex cones of generalized positive rational functions and the Nevanlinna-Pick interpolation. Linear Algebra Appl. **438**(10), 3949–3966 (2013)

[AJL13] D. Alpay, P.E.T. Jorgensen, I. Lewkowicz, Parametrizations of all wavelet filters: input-output and state-space. Sampl. Theory Signal Image Process. **12**(2–3), 159–188 (2013)

[AJS14] D. Alpay, P.E.T. Jorgensen, G. Salomon, On free stochastic processes and their derivatives. Stoch. Process. Appl. **124**(10), 3392–3411 (2014)

[AJV14] D. Alpay, P.E.T. Jorgensen, D. Volok, Relative reproducing kernel Hilbert spaces. Proc. Am. Math. Soc. **142**(11), 3889–3895 (2014)

[AJLM15] D. Alpay, P.E.T. Jorgensen, I. Lewkowicz, I. Martziano, Infinite product representations for kernels and iterations of functions, in *Recent Advances in Inverse Scattering, Schur Analysis and Stochastic Processes*. Operator Theory: Advances and Applications, vol. 244 (Birkhäuser/Springer, Cham, 2015), pp. 67–87

[AJK15] D. Alpay, P.E.T. Jorgensen, D.P. Kimsey, Moment problems in an infinite number of variables. Infin. Dimens. Anal. Quantum Probab. Relat. Top. **18**(4), 1550024 (2015)

[AJLV16] D. Alpay, P.E.T. Jorgensen, I. Lewkowicz, D. Volok, A new realization of rational functions, with applications to linear combination interpolation, the Cuntz relations and kernel decompositions. Complex Var. Elliptic Equ. **61**(1), 42–54 (2016)

© Springer International Publishing AG, part of Springer Nature 2018
S. Bezuglyi, P. E. T. Jorgensen, *Transfer Operators, Endomorphisms,*
and Measurable Partitions, Lecture Notes in Mathematics 2217,
https://doi.org/10.1007/978-3-319-92417-5

[ACKS16] D. Alpay, F. Colombo, D.P. Kimsey, I. Sabadini, The spectral theorem for unitary operators based on the S-spectrum. Milan J. Math. **84**(1), 41–61 (2016)

[AJL16] D. Alpay, P.E.T. Jorgensen, I. Lewkowicz, W-markov measures, transfer operators, wavelets and multiresolutions (2016). arXiv:1606.07692

[AR15] S.E. Arklint, E. Ruiz, Corners of Cuntz-Krieger algebras. Trans. Am. Math. Soc. **367**(11), 7595–7612 (2015)

[AM16] K. Arslan, V. Milousheva, Meridian surfaces of elliptic or hyperbolic type with pointwise 1-type Gauss map in Minkowski 4-space. Taiwan. J. Math. **20**(2), 311–332 (2016)

[BJMP05] L. Baggett, P. Jorgensen, K. Merrill, J. Packer, A non-MRA C^r frame wavelet with rapid decay. Acta Appl. Math. **89**(1–3), 251–270 (2006)

[BFMP09] L.W. Baggett, V. Furst, K.D. Merrill, J.A. Packer, Generalized filters, the low-pass condition, and connections to multiresolution analyses. J. Funct. Anal. **257**(9), 2760–2779 (2009)

[BLP$^+$10] L.W. Baggett, N.S. Larsen, J.A. Packer, I. Raeburn, A. Ramsay, Direct limits, multiresolution analyses, and wavelets. J. Funct. Anal. **258**(8), 2714–2738 (2010)

[BMPR12] L.W. Baggett, K.D. Merrill, J.A. Packer, A.B. Ramsay, Probability measures on solenoids corresponding to fractal wavelets. Trans. Am. Math. Soc. **364**(5), 2723–2748 (2012)

[BSV15] W. Bahsoun, J. Schmeling, S. Vaienti, On transfer operators and maps with random holes. Nonlinearity **28**(3), 713–727 (2015)

[BB05] M. Baillif, V. Baladi, Kneading determinants and spectra of transfer operators in higher dimensions: the isotropic case. Ergod. Theory Dyn. Syst. **25**(5), 1437–1470 (2005)

[Bal00] V. Baladi, *Positive Transfer Operators and Decay of Correlations*. Advanced Series in Nonlinear Dynamics, vol. 16 (World Scientific, River Edge, NJ, 2000)

[BER89] V. Baladi, J.-P. Eckmann, D. Ruelle, Resonances for intermittent systems. Nonlinearity **2**(1), 119–135 (1989)

[BJL96] V. Baladi, Y.P. Jiang, O.E. Lanford III, Transfer operators acting on Zygmund functions. Trans. Am. Math. Soc. **348**(4), 1599–1615 (1996)

[BRC16] K. Bandara, T. Rüberg, F. Cirak, Shape optimisation with multiresolution subdivision surfaces and immersed finite elements. Comput. Methods Appl. Mech. Eng. **300**, 510–539 (2016)

[BHS08] M.F. Barnsley, J.E. Hutchinson, Ö. Stenflo, V-variable fractals: fractals with partial self similarity. Adv. Math. **218**(6), 2051–2088 (2008)

[BHS12] M. Barnsley, J.E. Hutchinson, Ö. Stenflo, V-variable fractals: dimension results. Forum Math. **24**(3), 445–470 (2012)

[Bea91] A.F. Beardon, *Iteration of Rational Functions: Complex Analytic Dynamical Systems*. Graduate Texts in Mathematics, vol. 132 (Springer, New York, 1991)

[BCD16] B. Bektacs, E.Ö. Canfes, U. Dursun, On rotational surfaces in pseudo-Euclidean space $\mathbb{E}_T^4$ with pointwise 1-type Gauss map. Acta Univ. Apulensis Math. Inform. **45**, 43–59 (2016)

[Bén96] C. Bénéteau, A natural extension of a nonsingular endomorphism of a measure space. Rocky Mt. J. Math. **26**(4), 1261–1273 (1996)

[BG91] S. Bezuglyi, V. Golodets, Weak equivalence and the structures of cocycles of an ergodic automorphism. Publ. Res. Inst. Math. Sci. **27**(4), 577–625 (1991)

[BH14] S. Bezuglyi, D. Handelman, Measures on Cantor sets: the good, the ugly, the bad. Trans. Am. Math. Soc. **366**(12), 6247–6311 (2014)

[BJ15] S. Bezuglyi, P.E.T. Jorgensen, Representations of Cuntz-Krieger relations, dynamics on Bratteli diagrams, and path-space measures, in *Trends in Harmonic Analysis and Its Applications*. Contemporary Mathematics, vol. 650 (American Mathematical Society, Providence, RI, 2015), pp. 57–88

[BK16]   S. Bezuglyi, O. Karpel, Bratteli diagrams: structure, measures, dynamics, in *Dynamics and Numbers*. Contemporary Mathematics, vol. 669. (American Mathematical Society, Providence, RI, 2016), pp. 1–36

[BKMS10] S. Bezuglyi, J. Kwiatkowski, K. Medynets, B. Solomyak, Invariant measures on stationary Bratteli diagrams. Ergod. Theory Dyn. Syst. **30**(4), 973–1007 (2010)

[BKMS13] S. Bezuglyi, J. Kwiatkowski, K. Medynets, B. Solomyak, Finite rank Bratteli diagrams: structure of invariant measures. Trans. Am. Math. Soc. **365**(5), 2637–2679 (2013)

[BKLR15] M. Bischoff, Y. Kawahigashi, R. Longo, K.-H. Rehren, *Tensor Categories and Endomorphisms of von Neumann Algebras—With Applications to Quantum Field Theory*. Springer Briefs in Mathematical Physics, vol. 3 (Springer, Cham, 2015)

[Bog07]   V.I. Bogachev, *Measure Theory*, vols. I, II (Springer, Berlin, 2007)

[Bra72]   O. Bratteli, Inductive limits of finite dimensional C^*-algebras. Trans. Am. Math. Soc. **171**, 195–234 (1972)

[BJ97]   O. Bratteli, P.E.T. Jorgensen, Endomorphisms of $B(\mathcal{H})$. II. Finitely correlated states on $\mathcal{O}_n$. J. Funct. Anal. **145**(2), 323–373 (1997)

[BJ02]   O. Bratteli, P. Jorgensen, *Wavelets Through a Looking Glass: The World of the Spectrum*. Applied and Numerical Harmonic Analysis (Birkhäuser, Boston, MA, 2002)

[BK00]   O. Bratteli, A. Kishimoto, Homogeneity of the pure state space of the Cuntz algebra. J. Funct. Anal. **171**(2), 331–345 (2000)

[BEK93]   O. Bratteli, G.A. Elliott, A. Kishimoto, Quasi-product actions of a compact group on a C^*-algebra. J. Funct. Anal. **115**(2), 313–343 (1993)

[BJP96]   O. Bratteli, P.E.T. Jorgensen, G.L. Price, Endomorphisms of $B(\mathcal{H})$, in *Quantization, Nonlinear Partial Differential Equations, and Operator Algebra (Cambridge, MA, 1994)*. Proceedings of Symposia in Pure Mathematics, vol. 59 (American Mathematical Society, Providence, RI, 1996), pp. 93–138

[BH09]   H. Bruin, J. Hawkins, Rigidity of smooth one-sided Bernoulli endomorphisms. N. Y. J. Math. **15**, 451–483 (2009)

[CT16]   P.G. Casazza, J.C. Tremain, Consequences of the Marcus/Spielman/Srivastava solution of the Kadison-Singer problem, in *New Trends in Applied Harmonic Analysis*. Applied and Numerical Harmonic Analysis (Birkhäuser/Springer, Cham, 2016), pp. 191–213

[CL16]   X. Chao, Y. Lv, On the Gauss map of Weingarten hypersurfaces in hyperbolic spaces. Bull. Braz. Math. Soc. (N.S.) **47**(4), 1051–1069 (2016)

[CE77]   M.D. Choi, E.G. Effros, Injectivity and operator spaces. J. Funct. Anal. **24**(2), 156–209 (1977)

[CFS82]   I.P. Cornfeld, S.V. Fomin, Y.G. Sinaĭ, *Ergodic Theory*. Grundlehren der Mathematischen Wissenschaften [Fundamental Principles of Mathematical Sciences], vol. 245 (Springer, New York, 1982). Translated from the Russian by A. B. Sosinskiĭ

[DH94]   K.G. Dajani, J.M. Hawkins, Examples of natural extensions of nonsingular endomorphisms. Proc. Am. Math. Soc. **120**(4), 1211–1217 (1994)

[MdF16]   G.M. de Freitas, Submanifolds with homothetic Gauss map in codimension two. Geom. Dedicata **180**, 151–170 (2016)

[dlR06]   T. de la Rue, An introduction to joinings in ergodic theory. Discrete Contin. Dyn. Syst. **15**(1), 121–142 (2006)

[DZ09]   J. Ding, A. Zhou, *Nonnegative Matrices, Positive Operators, and Applications* (World Scientific, Hackensack, NJ, 2009)

[DJK94]   R. Dougherty, S. Jackson, A.S. Kechris, The structure of hyperfinite borel equivalence relations. Trans. Am. Math. Soc. **341**(1), 193–225 (1994)

[Dut02]   D.E. Dutkay, Harmonic analysis of signed Ruelle transfer operators. J. Math. Anal. Appl. **273**(2), 590–617 (2002)

[DJ06]   D.E. Dutkay, P.E.T. Jorgensen, Wavelets on fractals. Rev. Mat. Iberoam. **22**(1), 131–180 (2006)

[DJ07]   D.E. Dutkay, P.E.T. Jorgensen, Disintegration of projective measures. Proc. Am. Math. Soc. **135**(1), 169–179 (2007)

[DJ15]   D.E. Dutkay, P.E.T. Jorgensen, Representations of Cuntz algebras associated to quasi-stationary Markov measures. Ergod. Theory Dyn. Syst. **35**(7), 2080–2093 (2015)

[DR07]   D.E. Dutkay, K. Røysland, The algebra of harmonic functions for a matrix-valued transfer operator. J. Funct. Anal. **252**(2), 734–762 (2007)

[ES89]   S.J. Eigen, C.E. Silva, A structure theorem for n-to-1 endomorphisms and existence of nonrecurrent measures. J. Lond. Math. Soc. (2) **40**(3), 441–451 (1989)

[Exe03] R. Exel, A new look at the crossed-product of a C^*-algebra by an endomorphism. Ergod. Theory Dyn. Syst. **23**(6), 1733–1750 (2003)

[Fab87] R.C. Fabec, Induced group actions, representations and fibered skew product extensions. Trans. Am. Math. Soc. **301**(2), 489–513 (1987)

[Fab00] R.C. Fabec, *Fundamentals of Infinite Dimensional Representation Theory*. Chapman & Hall/CRC Monographs and Surveys in Pure and Applied Mathematics, vol. 114 (Chapman & Hall/CRC, Boca Raton, FL, 2000)

[FGKP16] C. Farsi, E. Gillaspy, S. Kang, J.A. Packer, Separable representations, KMS states, and wavelets for higher-rank graphs. J. Math. Anal. Appl. **434**(1), 241–270 (2016)

[Fed13] A.G. Fedotov, On the realization of the generalized solenoid as a hyperbolic attractor of sphere diffeomorphisms. Math. Notes **94**(5–6), 681–691 (2013). Translation of Mat. Zametki **94**(5), 733–744 (2013)

[FMCB⁺16] M. Focchi, G.A. Medrano-Cerda, T. Boaventura, M. Frigerio, C. Semini, J. Buchli, D.G. Caldwell, Robot impedance control and passivity analysis with inner torque and velocity feedback loops. Control Theory Technol. **14**(2), 97–112 (2016)

[GSSY16] D. Galicer, S. Saglietti, P. Shmerkin, A. Yavicoli, L^q dimensions and projections of random measures. Nonlinearity **29**(9), 2609–2640 (2016)

[GS16]   F.H. Ghane, A. Sarizadeh, Some stochastic properties of topological dynamics of semigroup actions. Topol. Appl. **204**, 112–120 (2016)

[Gla03] E. Glasner, *Ergodic Theory Via Joinings*. Mathematical Surveys and Monographs, vol. 101 (American Mathematical Society, Providence, RI, 2003)

[HW70]   P.R. Halmos, L.J. Wallen, Powers of partial isometries. J. Math. Mech. **19**, 657–663 (1970)

[Haw94] J.M. Hawkins, Amenable relations for endomorphisms. Trans. Am. Math. Soc. **343**(1), 169–191 (1994)

[HS91]   J.M. Hawkins, C.E. Silva, Noninvertible transformations admitting no absolutely continuous σ-finite invariant measure. Proc. Am. Math. Soc. **111**(2), 455–463 (1991)

[HŚ16]   K. Horbacz, M.Ślęczka, Law of large numbers for random dynamical systems. J. Stat. Phys. **162**(3), 671–684 (2016)

[Hut81] J.E. Hutchinson, Fractals and self-similarity. Indiana Univ. Math. J. **30**(5), 713–747 (1981)

[Hut96] J.E. Hutchinson, Elliptic systems, in *Instructional Workshop on Analysis and Geometry, Part I (Canberra, 1995)*. Proceedings of Centre for Mathematics and its Applications, vol. 34 (Australian National University Press, Canberra, 1996), pp. 111–120

[HR00]   J.E. Hutchinson, L. Rüschendorf, Selfsimilar fractals and selfsimilar random fractals, in *Fractal Geometry and Stochastics, II (Greifswald/Koserow, 1998)*. Progress in Probability., vol. 46 (Birkhäuser, Basel, 2000), pp. 109–123

[JMS16] P. Jaros, L. Maślanka, F. Strobin, Algorithms generating images of attractors of generalized iterated function systems. Numer. Algorithms **73**(2), 477–499 (2016)

[JLR16] Y.-Q. Ji, Z. Liu, S.-il Ri, Fixed point theorems of the iterated function systems. Commun. Math. Res. **32**(2), 142–150 (2016)

[Jon94] V.F.R. Jones, On a family of almost commuting endomorphisms. J. Funct. Anal. **122**(1), 84–90 (1994)

[Jor01] P.E.T. Jorgensen, Ruelle operators: functions which are harmonic with respect to a transfer operator. Mem. Am. Math. Soc. **152**(720), viii+60 (2001)

[Jor04] P.E.T. Jorgensen, Iterated function systems, representations, and Hilbert space. Int. J. Math. **15**(8), 813–832 (2004)

[JP11] P.E.T. Jorgensen, E.P.J. Pearse, Resistance boundaries of infinite networks, in *Random Walks, Boundaries and Spectra*. Progress in Probability, vol. 64 (Birkhäuser/Springer, Basel AG, Basel, 2011), pp. 111–142

[JP13] P.E.T. Jorgensen, E.P.J. Pearse, A discrete Gauss-Green identity for unbounded Laplace operators, and the transience of random walks. Isr. J. Math. **196**(1), 113–160 (2013)

[JS15] P.E.T. Jorgensen, M.-S. Song, Filters and matrix factorization. Sampl. Theory Signal Image Process. **14**(3), 171–197 (2015)

[JT15] P.E.T. Jorgensen, F. Tian, Infinite networks and variation of conductance functions in discrete Laplacians. J. Math. Phys. **56**(4), 043506, 27 (2015)

[JT16a] P.E.T. Jorgensen, F. Tian, Graph Laplacians and discrete reproducing kernel Hilbert spaces from restrictions. Stoch. Anal. Appl. **34**(4), 722–747 (2016)

[JT16b] P.E.T. Jorgensen, F. Tian, Positive definite kernels and boundary spaces. Adv. Oper. Theory **1**(1), 123–133 (2016)

[JT17] P. Jorgensen, F. Tian, Transfer operators, induced probability spaces, and random walk models. Markov Process. Relat. Fields **23**(2), 187–210 (2017)

[JPT15] P.E.T. Jorgensen, S. Pedersen, F. Tian, Spectral theory of multiple intervals. Trans. Am. Math. Soc. **367**(3), 1671–1735 (2015)

[Kak48] S. Kakutani, On equivalence of infinite product measures. Ann. Math. (2) **49**, 214–224 (1948)

[Kar59] S. Karlin, Positive operators. J. Math. Mech. **8**, 907–937 (1959)

[Kat07] M. Kato, Compactly supported framelets and the Ruelle operators, in *Applied Functional Analysis* (Yokohama Publ., Yokohama, 2007), pp. 177–191

[Kea72] M. Keane, Strongly mixing *g*-measures. Invent. Math. **16**, 309–324 (1972)

[Kec95] A.S. Kechris, *Classical Descriptive Set Theory*. Graduate Texts in Mathematics, vol. 156 (Springer, New York, 1995)

[KFB16] J.N. Kutz, X. Fu, S.L. Brunton, Multiresolution dynamic mode decomposition. SIAM J. Appl. Dyn. Syst. **15**(2), 713–735 (2016)

[LM94] A. Lasota, M.C. Mackey, *Chaos, Fractals, and Noise*. Applied Mathematical Sciences, vol. 97, 2nd edn. (Springer, New York, 1994). Stochastic aspects of dynamics

[LP13] F. Latrémolière, J.A. Packer, Noncommutative solenoids and their projective modules, in *Commutative and Noncommutative Harmonic Analysis and Applications*. Contemporary Mathematics, vol. 603 (American Mathematical Society, Providence, RI, 2013), pp. 35–53

[LP15] F. Latrémolière, J.A. Packer, Explicit construction of equivalence bimodules between noncommutative solenoids, in *Trends in Harmonic Analysis and Its Applications*. Contemporary Mathematics (American Mathematical Society, Providence, RI, 2015), pp. 111–140

[Lli15] J. Llibre, Brief survey on the topological entropy. Discrete Contin. Dyn. Syst. Ser. B **20**(10), 3363–3374 (2015)

[Lon89] R. Longo, Index of subfactors and statistics of quantum fields. I. Commun. Math. Phys. **126**(2), 217–247 (1989)

[Mai13] D. Maier, Realizations of rotations on *a*-adic solenoids. Math. Proc. R. Ir. Acad. **113A**(2), 131–141 (2013)

[MSS15] A.W. Marcus, D.A. Spielman, N. Srivastava, Interlacing families II: mixed characteristic polynomials and the Kadison-Singer problem. Ann. Math. (2) **182**(1), 327–350 (2015)

[Mat17]   K. Matsumoto, Uniformly continuous orbit equivalence of Markov shifts and gauge actions on Cuntz–Krieger algebras. Proc. Am. Math. Soc. **145**(3), 1131–1140 (2017)

[Mau95]   R.D. Mauldin, Infinite iterated function systems: theory and applications, in *Fractal Geometry and Stochastics (Finsterbergen, 1994)*. Progress in Probability, vol. 37 (Birkhäuser, Basel, 1995), pp. 91–110

[MU10]   V. Mayer, M. Urbański, Thermodynamical formalism and multifractal analysis for meromorphic functions of finite order. Mem. Am. Math. Soc. **203**(954), vi+107 (2010)

[MU15]   V. Mayer, M. Urbański, Countable alphabet random subshifts of finite type with weakly positive transfer operator. J. Stat. Phys. **160**(5), 1405–1431 (2015)

[Nel69]   E. Nelson, *Topics in Dynamics. I: Flows*. Mathematical Notes (Princeton University Press/University of Tokyo Press, Princeton, NJ/Tokyo, 1969)

[NR82]   F. Nicolò, C. Radin, A first-order phase transition between crystal phases in the shift model. J. Stat. Phys. **28**(3), 473–478 (1982)

[Par69]   W. Parry, *Entropy and Generators in Ergodic Theory* (W. A. Benjamin, New York-Amsterdam, 1969)

[PW72]   W. Parry, P. Walters, Endomorphisms of a Lebesgue space. Bull. Am. Math. Soc. **78**, 272–276 (1972)

[Pow99]   R.T. Powers, Induction of semigroups of endomorphisms of $\mathcal{B}(\mathcal{H})$ from completely positive semigroups of $(n \times n)$ matrix algebras. Inter. J. Math. **10**(7), 773–790 (1999)

[PP93]   R.T. Powers, G.L. Price, Binary shifts on the hyperfinite II_1 factor, in *Representation Theory of Groups and Algebras*. Contemporary Mathematics (American Mathematical Society, Providence, RI, 1993), pp. 453–464

[PU10]   F. Przytycki, M. Urbański, *Conformal Fractals: Ergodic Theory Methods*. London Mathematical Society Lecture Note Series, vol. 371 (Cambridge University Press, Cambridge, 2010)

[Rad99]   C. Radin, *Miles of Tiles*. Student Mathematical Library, vol. 1 (American Mathematical Society, Providence, RI, 1999)

[Ren87]   J. Renault, Représentation des produits croisés d'algèbres de groupoïdes. J. Operator Theory **18**(1), 67–97 (1987)

[Rén57]   A. Rényi, Representations for real numbers and their ergodic properties. Acta Math. Acad. Sci. Hung. **8**, 477–493 (1957)

[RG16]   F.O. Reveles-Gurrola, Homeomorphisms of a solenoid isotopic to the identity and its second cohomology groups. C. R. Math. Acad. Sci. Paris **354**(9), 879–886 (2016)

[Roh49b] V.A. Rohlin, On the fundamental ideas of measure theory. Mat. Sbornik N.S. **25**(67), 107–150 (1949)

[Roh49a] V.A. Rohlin, Selected topics from the metric theory of dynamical systems. Uspehi Matem. Nauk (N.S.) **4**(2(30)), 57–128 (1949)

[Roh61]   V.A. Rohlin, Exact endomorphisms of a Lebesgue space. Izv. Akad. Nauk SSSR Ser. Mat. **25**, 499–530 (1961)

[Rud90]   D.J. Rudolph, *Fundamentals of Measurable Dynamics: Ergodic Theory on Lebesgue Spaces*. Oxford Science Publications (Clarendon Press, Oxford University Press, New York, 1990)

[Rue78]   D. Ruelle, *Thermodynamic Formalism*. Encyclopedia of Mathematics and its Applications, vol. 5 (Addison-Wesley, Reading, MA, 1978). The mathematical structures of classical equilibrium statistical mechanics, With a foreword by Giovanni Gallavotti and Gian-Carlo Rota

[Rue89]   D. Ruelle, The thermodynamic formalism for expanding maps. Commun. Math. Phys. **125**(2), 239–262 (1989)

[Rue92]   D. Ruelle, Thermodynamic formalism for maps satisfying positive expansiveness and specification. Nonlinearity **5**(6), 1223–1236 (1992)

[Rue02]   D. Ruelle, Dynamical zeta functions and transfer operators. Not. Am. Math. Soc. **49**(8), 887–895 (2002)

[Rug16]   H.H. Rugh, The Milnor-Thurston determinant and the Ruelle transfer operator. Commun. Math. Phys. **342**(2), 603–614 (2016)

[SW17]   Y. Shiozawa, J. Wang, Rate functions for symmetric Markov processes via heat kernel. Potential Anal. **46**(1), 23–53 (2017)

[Sil88]   C.E. Silva, On μ-recurrent nonsingular endomorphisms. Isr. J. Math. **61**(1), 1–13 (1988)

[Sil13]   S. Silvestrov, Dynamics, wavelets, commutants and transfer operators satisfying crossed product type commutation relations, in *Operator Algebra and Dynamics*. Springer Proceedings of Mathematics & Statistics, vol. 58 (Springer, Heidelberg, 2013), pp. 273–293

[Sim12]   D. Simmons, Conditional measures and conditional expectation; Rohlin's disintegration theorem. Discrete Contin. Dyn. Syst. **32**(7), 2565–2582 (2012)

[Sto12]   L. Stoyanov, Regular decay of ball diameters and spectra of Ruelle operators for contact Anosov flows. Proc. Am. Math. Soc. **140**(10), 3463–3478 (2012)

[Sto13]   L. Stoyanov, Ruelle operators and decay of correlations for contact Anosov flows. C. R. Math. Acad. Sci. Paris **351**(17–18), 669–672 (2013)

[SG16]   A.J. Suarez, S. Ghosal, Bayesian clustering of functional data using local features. Bayesian Anal. **11**(1), 71–98 (2016)

[Sur16]   C. Sureson, Π_1^1-Martin-Löf random reals as measures of natural open sets. Theor. Comput. Sci. **653**, 26–41 (2016)

[SUZ13]   T. Szarek, M. Urbański, A. Zdunik, Continuity of Hausdorff measure for conformal dynamical systems. Discrete Contin. Dyn. Syst. **33**(10), 4647–4692 (2013)

[Sze17]   Z.S. Szewczak, Berry-Esséen theorem for sample quantiles of asymptotically uncorrelated non reversible Markov chains. Commun. Stat. Theory Methods **46**(8), 3985–4003 (2017)

[Ver81]   A.M. Vershik, Uniform algebraic approximation of shift and multiplication operators. Dokl. Akad. Nauk SSSR **259**(3), 526–529 (1981)

[Ver94]   A.M. Vershik, Theory of decreasing sequences of measurable partitions. Algebra i Analiz **6**(4), 1–68 (1994)

[Ver00]   A.M. Vershik, Dynamic theory of growth in groups: entropy, boundaries, examples. Uspekhi Mat. Nauk **55**(4(334)), 59–128 (2000)

[Ver01]   A.M. Vershik, V. A. Rokhlin and the modern theory of measurable partitions, in *Topology, Ergodic Theory, Real Algebraic Geometry*. American Mathematical Society, Translations: Series 2, vol. 202 (American Mathematical Society, Providence, RI, 2001), pp. 11–20

[Ver05]   A.M. Vershik, Polymorphisms, Markov processes, and quasi-similarity. Discrete Contin. Dyn. Syst. **13**(5), 1305–1324 (2005)

[VF85]   A.M. Vershik, A.L. Fëdorov, Trajectory theory, in *Current Problems in Mathematics. Newest Results*, vol. 26. Itogi Nauki i Tekhniki (Akad. Nauk SSSR, Vsesoyuz. Inst. Nauchn. i Tekhn. Inform., Moscow, 1985), pp. 171–211, 260

[Wol48]   H.O.A. Wold, On prediction in stationary time series. Ann. Math. Stat. **19**, 558–567 (1948)

[Wol51]   H.O.A. Wold, Stationary time series. Trabajos Estadística **2**, 3–74 (1951)

[Wol54]   H. Wold, *A Study in the Analysis of Stationary Time Series*, 2nd edn. (Almqvist and Wiksell, Stockholm, 1954). With an appendix by Peter Whittle

[YL16]   Y. Yao, W. Li, Generating iterated function systems for the Vicsek snowflake and the Koch curve. Am. Math. Mon. **123**(7), 716–721 (2016)

[YLZ99]   D. Yan, X. Liu, W. Zhu, A study of Mandelbrot and Julia sets generated from a general complex cubic iteration. Fractals **7**(4), 433–437 (1999)

[YZL13] R. Ye, Y. Zou, J. Lu, Chaotic dynamical systems on fractals and their applica-
         tions to image encryption, in *Recent Advances in Applied Nonlinear Dynamics
         with Numerical Analysis*. Interdisciplinary Mathematical Sciences, vol. 15 (World
         Scientific, Hackensack, NJ, 2013), pp. 279–304
[ZJ15]   Z. Zhang, P.E.T. Jorgensen, Modulated Haar wavelet analysis of climatic back-
         ground noise. Acta Appl. Math. **140**, 71–93 (2015)

Index

automorphic factor, 89, 91
automorphism, 14, 18, 21, 89, 91

Bratteli, 148
Bratteli diagram, 148

coboundary
  σ-coboundary, 37, 64, 65, 72, 109, 116
  R-coboundary, 37
cocycle, 37
coding map, 137
composition operator, 24, 57
conditional expectation, 32, 102, 103, 114, 141

density, 113
Doob transform, 35

encoding space, 136
endomorphism, 14
endomorphism
  exact, 17
  piecewise monotone, 119
  Borel, 6
  conservative, 16
  countable-to-one, 20
  ergodic, 16
  finite-to-one, 133
  Gauss, 126

measure preserving, 15
  non-singular, 15
equivalent pairs, 93
ergodic components, 18
ergodic decomposition, 49

factor, 89
factor map, 88
Frobenius-Perron operator, 8

Gauss map, 126
generalized conditional expectation, 32

harmonic function, 24, 34, 35, 74, 104, 109, 116, 146

inverse branch, 133
iterated function system (IFS), 120

Jacobian, 148
joining, 58

Koopman operator, 24

Markov, 2
Markov
  property, 58

© Springer International Publishing AG, part of Springer Nature 2018
S. Bezuglyi, P. E. T. Jorgensen, *Transfer Operators, Endomorphisms,
and Measurable Partitions*, Lecture Notes in Mathematics 2217,
https://doi.org/10.1007/978-3-319-92417-5

LECTURE NOTES IN MATHEMATICS

 Springer

Editors in Chief: J.-M. Morel, B. Teissier;

Editorial Policy

1. Lecture Notes aim to report new developments in all areas of mathematics and their applications – quickly, informally and at a high level. Mathematical texts analysing new developments in modelling and numerical simulation are welcome.

   Manuscripts should be reasonably self-contained and rounded off. Thus they may, and often will, present not only results of the author but also related work by other people. They may be based on specialised lecture courses. Furthermore, the manuscripts should provide sufficient motivation, examples and applications. This clearly distinguishes Lecture Notes from journal articles or technical reports which normally are very concise. Articles intended for a journal but too long to be accepted by most journals, usually do not have this "lecture notes" character. For similar reasons it is unusual for doctoral theses to be accepted for the Lecture Notes series, though habilitation theses may be appropriate.

2. Besides monographs, multi-author manuscripts resulting from SUMMER SCHOOLS or similar INTENSIVE COURSES are welcome, provided their objective was held to present an active mathematical topic to an audience at the beginning or intermediate graduate level (a list of participants should be provided).

   The resulting manuscript should not be just a collection of course notes, but should require advance planning and coordination among the main lecturers. The subject matter should dictate the structure of the book. This structure should be motivated and explained in a scientific introduction, and the notation, references, index and formulation of results should be, if possible, unified by the editors. Each contribution should have an abstract and an introduction referring to the other contributions. In other words, more preparatory work must go into a multi-authored volume than simply assembling a disparate collection of papers, communicated at the event.

3. Manuscripts should be submitted either online at www.editorialmanager.com/lnm to Springer's mathematics editorial in Heidelberg, or electronically to one of the series editors. Authors should be aware that incomplete or insufficiently close-to-final manuscripts almost always result in longer refereeing times and nevertheless unclear referees' recommendations, making further refereeing of a final draft necessary. The strict minimum amount of material that will be considered should include a detailed outline describing the planned contents of each chapter, a bibliography and several sample chapters. Parallel submission of a manuscript to another publisher while under consideration for LNM is not acceptable and can lead to rejection.

4. In general, **monographs** will be sent out to at least 2 external referees for evaluation.

   A final decision to publish can be made only on the basis of the complete manuscript, however a refereeing process leading to a preliminary decision can be based on a pre-final or incomplete manuscript.

   Volume Editors of **multi-author works** are expected to arrange for the refereeing, to the usual scientific standards, of the individual contributions. If the resulting reports can be forwarded to the LNM Editorial Board, this is very helpful. If no reports are forwarded or if other questions remain unclear in respect of homogeneity etc, the series editors may wish to consult external referees for an overall evaluation of the volume.

5. Manuscripts should in general be submitted in English. Final manuscripts should contain at least 100 pages of mathematical text and should always include

 – a table of contents;
 – an informative introduction, with adequate motivation and perhaps some historical remarks: it should be accessible to a reader not intimately familiar with the topic treated;
 – a subject index: as a rule this is genuinely helpful for the reader.
 – For evaluation purposes, manuscripts should be submitted as pdf files.

6. Careful preparation of the manuscripts will help keep production time short besides ensuring satisfactory appearance of the finished book in print and online. After acceptance of the manuscript authors will be asked to prepare the final LaTeX source files (see LaTeX templates online: https://www.springer.com/gb/authors-editors/book-authors-editors/manuscriptpreparation/5636) plus the corresponding pdf- or zipped ps-file. The LaTeX source files are essential for producing the full-text online version of the book, see http://link.springer.com/bookseries/304 for the existing online volumes of LNM). The technical production of a Lecture Notes volume takes approximately 12 weeks. Additional instructions, if necessary, are available on request from lnm@springer.com.

7. Authors receive a total of 30 free copies of their volume and free access to their book on SpringerLink, but no royalties. They are entitled to a discount of 33.3 % on the price of Springer books purchased for their personal use, if ordering directly from Springer.

8. Commitment to publish is made by a *Publishing Agreement*; contributing authors of multiauthor books are requested to sign a *Consent to Publish form*. Springer-Verlag registers the copyright for each volume. Authors are free to reuse material contained in their LNM volumes in later publications: a brief written (or e-mail) request for formal permission is sufficient.

Addresses:
Professor Jean-Michel Morel, CMLA, École Normale Supérieure de Cachan, France
E-mail: moreljeanmichel@gmail.com

Professor Bernard Teissier, Equipe Géométrie et Dynamique,
Institut de Mathématiques de Jussieu – Paris Rive Gauche, Paris, France
E-mail: bernard.teissier@imj-prg.fr

Springer: Ute McCrory, Mathematics, Heidelberg, Germany,
E-mail: lnm@springer.com

Printed in the United States
By Bookmasters